Draft:Free Energy does not Exist

 The legitimacy of this research has been questioned. See the discussion page for more information.

How much Energy Does it Take to Operate Reactance (also known as Reactive Power)?

Answer ...

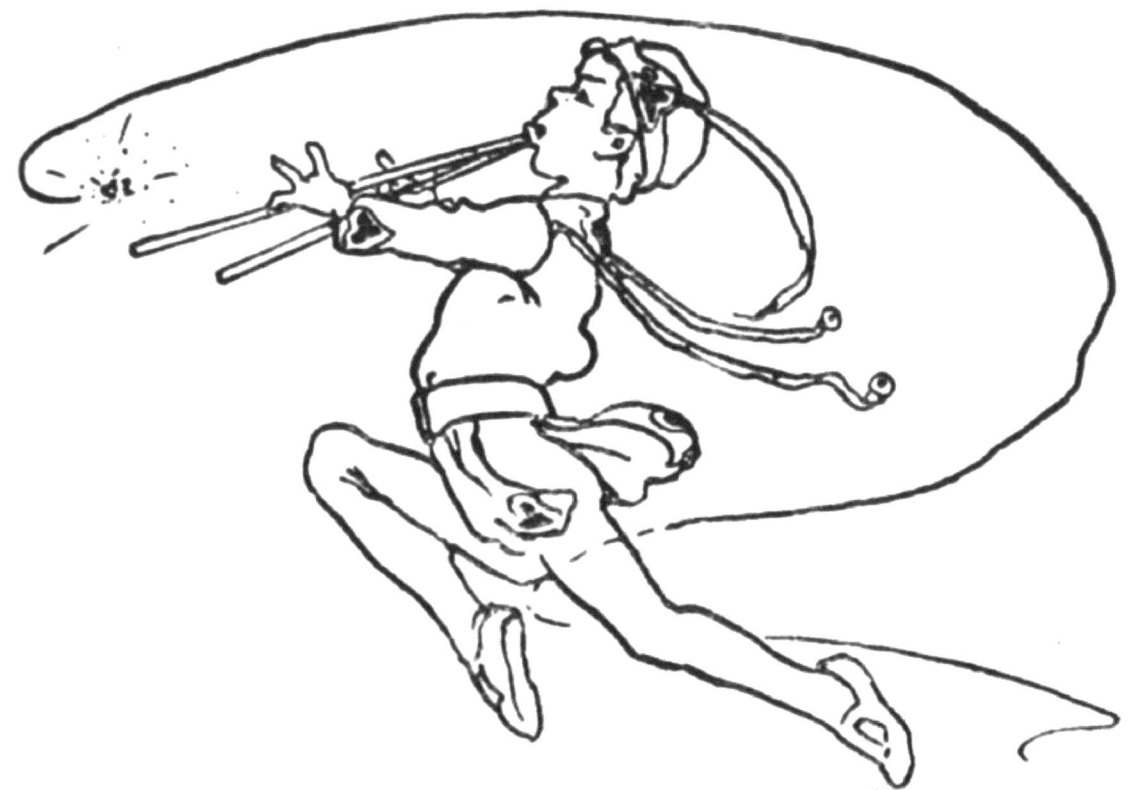

Relativity Explains Free Energy (https://www.youtube.com/watch?v=bOCOhrAsV8M) – **a video monologue on YouTube, 55⅔ min**

Or ...

The relativity of physics completely provides for the proof of free energy! (https://vinyasi.podbean.com/e/the-relativity-of-physics-completely-provides-for-the-proof-of-free-energy/) – an audio podcast, 52 min

> The physicist has borrowed the concept of reactance (but not for long if I can help it!), both electrical and magnetic, by misappropriating it to themselves by putting a new spin on it by calling it by a new name of, "relativity", giving it a fresh perspective. It is high time we acknowledge this contribution to the field of electrical and magnetic engineering by taking back what has not originally belonged to the field of physics.
>
> Relativity is the mechanism by which we measure energy in terms of reactance. Thus, reactance alters energy and removes energy from out of the static realm of solid objects (energetically tied down to their material expressions) and redefines energy upon the perspective which reactance contributes to our measurement of energy. Relativity makes it possible for energy to become measurable and quantifiable exclusively because of, and in terms of, reactance. Energy cannot be separated from reactance.

Reactance is the elusive field of the "ether" which scientists, of a bygone era of over a century ago, sought to justify an absolute frame of reference. They were wrong in pursuing an absolute frame of reference since it does not exist. This is why they never found it.

Reactance is not real; it's not physical. Only energy is of the material world while reactance has to be inferred to exist since it is structured out of the square root of negative one. And the square root of negative one is a manmade fiction cooked up to explain the elusive world of reactance in the context of, and as a function of, relativity.

Yet, this elusive reactance is the predicate for our perception, and -thus- our measurement, of energy. For, without the immaterial manmade concept of time (for example) a duration could not become associated with a physical mile to give us the speedometer to measure how fast we accelerate to pass the vehicle in front of us whenever driving on the highway.

And, Time gives us a generic factor of relativistic reactance called: duration. For, without duration, we could not be honestly and accurately billed by the electric company for our energy usage.

But, Time, as a duration, is only one factor of relativistic reactance and a generic factor at that. There are, also, the specific factors of electrical and magnetic reactances and I've merely studied the factors which are specific to electrical reactance ignoring (for these past six years) any study of the factors of magnetic reactance.

Yet, These factors of electrical reactance have been sufficient for me to program the circuit (of a simulation) to revise the circuit's perspective of energy over a period of time (a duration) in which the circuit's perception becomes altered in a gradual or sudden manner such that: the energy which I feed it, or fed it, is redefined due to the circuit's manipulation of energy via the factors of electrical reactance. This manipulation is capable of altering the circuit's perception of energy over a period of time so that the circuit can export 100,000 watts instead of something less than what I fed it (which, normally, amounts to pico watts or nano watts or femto watts of input).

The overunity circuit which I design will initially exhibit a circulation of energy inside of itself which is less than what I have fed it, but will gradually (or suddenly and explosively) alter its perception of its internal energy so that, over time, that perception becomes redefined so as to exhibit a greater quantity of energy than what the circuit perceived was initially circulating inside of itself. This transformation of perception is afforded us due to the relativistic nature of reactance.

It takes an indeterminate quantity of energy to operate reactance. It takes some energy, but the actual required amount may vary or be irrelevant due to the parametric property of certain types of reactance.

But just because the quantity of reactance may not matter does not make it possible to do away with it altogether unless we make up the difference with an unlimited supply of energy (and money to pay for that energy) to operate a non-reactive, or semi-reactive, appliance.

If we make an analogy between energetic water flowing within a reactive medium such as through a conduit (pipe) representing current flowing through a conductor (possessing inductance), then either an infinite amount of time or an infinitely large pipe will single-handedly be capable of conducting an unlimited quantity of water through that pipe.

And, since time is regulated by the frequency of oscillations and also regulates the phase relations between voltage and current, time is another factor of reactance besides capacitance and inductance. Thus, we have four factors to regulate the pumping of energy against a gradient of impedance – not by fighting that gradient, but – by effortlessly reversing current under certain conditions.

And, since the size of a pipe can be enlarged, its equivalency of inductance within a conductive medium may also be enlarged to any size desired to accommodate any quantity of water which is carried within a pipe or any quantity of current within a conductor in any shortness of transit-time.

Think of a heat pump. It pumps water against a gradient of greater heat to increase that heat. Thus, it should (theoretically at least) get harder and harder to pump against an ever-increasing gradient of escalating heat, yes?

That's what happens if we ignore the power of reactance and focus all of our effort upon energetically pumping energy against a gradient, namely: against impedance.

But with the reversal of current, the opposite happens. It gets easier and easier to reactively pump energy against an ever-increasing voltage due to the constant acceleration of pumping action brought about by escalating reactances.

There are four reactances that we may manipulate to achieve these results. They are: mutual inductance, self-inductance, mutual capacitance and self-capacitance.

The mutual varieties of inductance and capacitance are an interesting pair of reactances, for they are inverses of each other. In other words, whenever mutual inductance is high, mutual capacitance is low. And, whenever mutual capacitance is high, mutual inductance is low.

If they appear at all (and they, usually, always do), then they always appear together at the same time. They are never capable of entirely excluding each other. It is impossible for a circuit to possess one without also possessing the other.

These two properties of mutual reactance trump the self-referral versions in that, for some uncanny reason, mutual reactance may escalate, or diminish, over time even though self-reactance cannot. Mutual reactances are, thus: parametric.

Go, figure!

Power is the rate of Energy Usage. Power can also imply Efficiency of Energy Usage and can be less than, or more than, Unity.

Reactance always affects Power to one extent or another.

Time is not the only factor which can modify the Reactance of Energy Usage. All of the factors of Electrical and Magnetic Reactance can also modify Reactive Power, such as: Frequency, Duration, the Phase Relation between Voltage and Current, Capacitance and Inductance (just to name a few of the numerous factors of *Reactive* Power).

It may take a minimum quantity of Energy to accomplish a task, but this is superseded by having sufficient Power (both Real and Reactive) to get the job done. Thus, don't allow yourself to be shocked when any overunity circuit exhibits a Wattage of Output exceeding its Wattage of Input.

Energy versus Power

Energy versus Power (https://energyeducation.ca/encyclopedia/Energy_vs _power)

Free Energy is the presumption that Energy is *infinitely available*[a] due to a potentially limitless supply of Reactance (both, Magnetic and Electric) versus **Power** which is (at the very least) the method of regulating the rate at which Energy is used per unit of Time. Consequently, **Free Energy** has more to do with the non-accountability of **Power** than it has to do with **Energy** since Reactant Energy is assumed to be unstable when its frame of reference is altered from its default condition (established by Nature) through the interventions of the eminent domain of the electric utility grid or by perpetual motion machines which are assumed to be isolated from their environment. This charade makes the **Reactant** byproduct of **Energy** *appear to be limitless* for all intents and purposes so long as there is always sufficient wiggle-room for the alteration of perspective which Reactance affords us. This is what makes Energy: **"Free."** It is the boundless Relativity of Reactance.

The **reversal of current**, which is brought about by various reactances of one kind or another, is the method by which energy may be shuffled around the Universe *against* voltage gradients and *effectively uphill*. This overrides the diffusion of energy which would have resulted in entropy and the elimination of voltage differences between any two points in space. Without these voltage differences, we are powerless to make use of energy no matter how much energy may reside within our Universe. Thus, the *negation of current* allows us to supersede the side-effects of the natural order of Cosmic Law as our Universe winds down from exhaustion (as described by physics) and safeguard our material existence *despite this natural tendency* (of an entropic Universe) and *because of this safeguard* of negative impedance (the *reversal of current*).

What Free Energy is Not ...

Free Energy is *not* a high Quality factor, nor is it an **infinite** Quality factor. In fact, it possesses an **undefined** or an **indeterminate** Q factor in most situations! Why? Because the *net power loss*, after adding whatever overunity was effectively gained, is zero.

When net power loss = 0, then ...

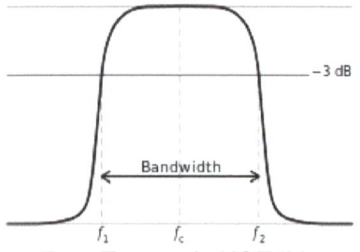

Free Energy is NOT this:
A graph of a conventional filter's gain magnitude, illustrating the concept of −3 dB at a voltage gain of 0.707 or half-power bandwidth. The frequency axis of this symbolic diagram can be linear or logarithmically scaled.

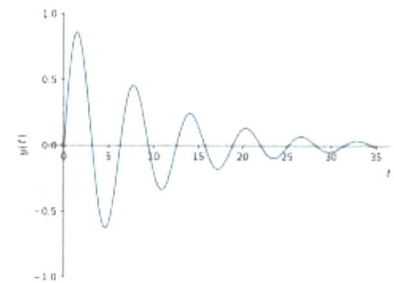

Free Energy is NOT this:
A damped oscillation. A low Q factor – about 5 here – means the oscillation dies out rapidly.

$$Q \stackrel{\text{def}}{=} 2\pi \times \frac{\text{energy stored}}{\text{energy dissipated per cycle}} = 2\pi f_r \times \frac{\text{energy stored}}{\text{net power loss}} = \text{Undefined, Indeterminate}$$

... because, zero in the denominator of any fraction is, *at worst,* not-a-number! At best, zero in the denominator might render as infinity within *real or complex projective geometry* if the numerator is not also zero.[1] But, this latter possibility is not the case.

A normal graph of Q factor (shown on the left) cannot occur with **Free Energy**, because damping (shown on the right) is impossible (by the rules of multiplication and division) whenever the net power loss is zero!

An Example of What Free Energy Is ...

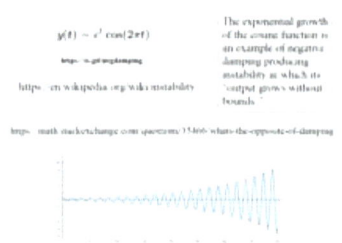

Free Energy is this:
The exponential growth of the cosine function (https://www.wolframalpha.com/input?i2d=true&i=graph+Power%5Be%2C%5C%2840%29t%5C%2841%29%5D+cos+%5C%2840%292%CF%80t%5C%2841%29) is an example of negative damping (https://math.stackexchange.com/questions/35466/whats-the-opposite-of-damping) producing instability in which its "output grows without bounds." $y(t) = e^t \cos(2\pi t)$

Here is an example (https://www.wolframalpha.com/input?i2d=true&i=graph+Power%5Be%2C%5C%2840%29t%5C%2841%29%5D+cos+%5C%2840%292%CF%80t%5C%2841%29) (on the left) of what **Free Energy** is: negative damping (https://math.stackexchange.com/questions/35466/whats-the-opposite-of-damping) ...

... during certain disturbances involving **generation loss** in the Northern region of India's power grid. – Excerpted from: *Low Frequency Oscillations in Indian Grid; VII. Conclusion.* Figure #2 on PDF page 3, and Figures #6 and #7 on PDF page 4.[2]

I suspect that the reason why the electric utility grid of Northern India generated power (seemingly, from within itself) when some its generators were offline was due to negative damping of its reactive storage elements, such as: within its capacitors, its inductors, and its conductive transmission lines. The capacitance of its grid performed the actual generation of reactive negative damping while its inductance converted this reactive format into real power which could be measured and taken notice of its eruption by those people who were, and are, responsible for managing that grid.

Empty space, and the areas of space which are filled with matter, acts as a dielectric medium with access to all of the electrical energy of all of the matter within the entire universe. All of matter functionally serves as a collective set of conductive plates comprising a Cosmic Capacitor while the dielectric of empty space stores energy within this Cosmic Capacitance. Unlike an inductor, acting as a conductor, a capacitor does not transmit energy from the plate on one side of its dielectric to the plate on the other side. A capacitor merely *stores* energy in its dielectric medium which is sandwiched between its conductive plates. This storage is evenly distributed throughout its dielectric material. In the case of the Universe, every particle of matter collectively contributes to the multiple plates of our Cosmic Capacitance. And the entirety of empty space within the whole Universe has direct access to the energy which is conducted to empty space by all of the matter which resides within empty space.

So, ...

Where does **Free Energy** come from? If a *free energy circuit* is producing overunity, where does it get its extra energy from?

Simulators don't give us the answer. Simulations of free energy circuits suggest that free energy circuits will either *steal* energy from out of nearby power lines,[3] or else will miraculously manufacture free energy (over time) from out of themselves as if they had an unlimited, 'secret' supply.

If you wish to play this virtual mind game in a conservative fashion, then I would suggest that all free energy is *stolen* from the environment via empty space. In other words, like it or not, all of material existence supplies energy for each and every free energy circuit whether or not we, or anyone else, wants to share our energy *for free* with anyone else is a moot point since *all energy*, spread across the entire Universe, is available for use by anyone who is willing and able to steal it in this fashion.

This does not violate the Conservation of Energy since the Conservation of Energy is a fictional concept which describes an ideal case of the presumed isolation of an energetic system from its surrounding environment. This presumption is, actually, less than ideal since it presupposes an extreme lack of efficiency in which that system is purposely engineered to be so deficient as to totally ignore its environment at the cost of that system demanding, from us, a perpetual subservience for keeping that energetic system alive with additional energy continuously provided **by us** if we want that energetic system to function at all!

This stinks, to me, of being the result of the mass institution of a policy than it being the byproduct of some law of physics, for it reinforces slavery to our machines.

But if we consider that *all* electrical circuits are open to their environment to some extent, then Newton's Third Law[4] fails to predict the Lorentz force and overlooks *free energy*.[5]

No Longer Groping for the Truth

I used to feel the need to judge and blame physicists for the lies and theft which they have protracted against electrical engineers and inventors in the name of science and my disdain for politics (namely, the Patent Office in every country) inaugurating these atrocities in the name of good will and common sense! But, that was due to my having to grope my way through the numerous lies to get to the single-most important truth.

There are many truths to this saga. But some truths are more relevant than other truths. Blame and the judgment of sinners is not one of them.

The most significant truth is to congratulate physicists for having contributed to our pursuit of knowledge regardless of how they have gone about it. They have given us "relativity" which helps to explain why energy can *appear* or *disappear* so convincingly that we choose to ignore these anomalies claiming, instead, that there is some fault somewhere else giving us, what we presume to be, false information.

In truth, relativity explains the non-creation and the non-destruction of energy when we are faced with its unaccountable (ie, non-thermodynamic) appearance and disappearance by providing us with a shift in our perception of energy. This shift is so convincing, that even our circuits are convinced!

This shift occurs due to all of our measurements of energy are founded upon power and what impact power always has upon our measurement – and perception – of energy. This foundation is due to the presence of reactive power since real power plays no role in the modification of our measurement of energy. Real power "conserves" our perception of energy against its modification while reactive power does not, and cannot, prevent this alteration of perception to occur.

Since reactive power is a fiction born of another manmade fiction known as: the square root of negative one, no one has the guts to admit that half of electrical engineering is predicated upon a belief in a fiction! Meanwhile, physicists conveniently ignore this salient fact.

But, relativity allows reactance to alter the magnitude of energy so convincingly and so benignly and so seamlessly that we tend to ignore this alteration when it happens or, else, explain it away with some trite excuse.

There is no other way to explain "free energy." Reactance can sometimes act like a pump, whenever the over-reactive, negative impedance of current reversal occurs (due to Foster's reactance theorem). This current reversal is capable of pumping energy *against* gradients of positive impedance and simple resistance.

And if a manmade device (such as: the electric utility grid) should get in the way of this pumping action, then so be it. The energy of this manmade device will be running the risk of having its energy, either: stolen from itself, or -else- it will have energy added to itself, due to the over-reactance (and the sustained reversal of current) of the free energy device.

Thank God for relativity!

Teaser

Question posed on Quora ...[6]

How can I make money using the science of energy?

My answer ...

This probably does not answer your question in the manner to which you had intended, but an alternative answer is to "save money which would have been spent had we not taken advantage of the science of energy."

So, this possibility could make you richer, not by bringing in more income, but by saving on the expense of energy. This is precisely what free energy is all about.

Free energy is not about discovering some hidden source of energy so much as it is in making use of potential energies which are largely overlooked, such as the potentialities of capacitance, inductance, phase relations and frequency (the features of electrical reactance).

Students of electrical engineering are programmed to believe that the first two potentialities of electrical reactance, namely: of capacitance and inductance, are merely temporary methods of energy storage. They are not seen as potential contributors to the expansion of energy because energy is always treated as if it were a physical object, or equivalent to same, such as: a cord of wood which we burn in the fireplace.

Wood is seen as being a physical object whose material existence, in turn, is seen as being a conservable quantity subject to the limitations of thermodynamic entropy, meaning that: systems of energy will eventually diminish to zero values of energy if not continuously pumped from outside of themselves with new reserves of energy to make up for whatever energy has already been spent. This is equivalent to the analogy of solid wood which, once spent/burnt, must be replaced.

This is not also true of electrical energy (and -possibly- also not true of energy in general).

We are told that electrical energy is a thing which is confined to a quantity. This is what the Conservation of Energy (by the study of physics) asserts without ambiguity. But in deference to the non-ambiguity assertion of physics, the reactance of a circuit which responds to that energy is not a quantity so much as it is a ratio of expansion or contraction of what we perceive to be the quantity of energy. So, ultimately, quantities of energy are not real, but are illusory since these quantities are not the foundation of energy. In other words, due to the intervention of electrical reactance, our quantification of energy is not a premise, but is dependent upon our _perception of quantity_ to form a premise of energy quantification.

Hence, energy may contract or expand and, thus, vary our perception of its quantity simply by applying the principles of electrical reactance in the form of a free energy circuit so that whatever energy enters this type of circuit will not _appear to equal_ the energy which exits it. This perceptual measurement of ours is fallacious since energy is not a thing so much as it is the perception of an assumption that _we are measuring a thing_ when, in fact, we are measuring how long it will take for the circuit to die from lack of energy. _Time_ is what we manipulate within a free energy circuit, not energy. And it is this manipulation of time which alters our perception of the quantity of energy which is available under reactive storage.[7]

In other words, if we can prolong a circuit's lifespan by recycling X seconds of energy (using capacitors and inductors), and recycle this quantity of energy once every $X \div 2$ seconds, then with the passage in time of every $X \div 2$ seconds, the duration of energy expenditure stored within these capacitors and inductors will _appear to double_. And if the total duration of the expenditure of energy, within this type of circuit, depletes its duration of storage by anything less than the amount which is recycled per unit of $X \div 2$ seconds, then our perception of this circuit's active lifespan will continue to increase at an exponential rate without limit until this circuit dies from a different sort of death, namely: death by frying itself from all of the heat it will generate plus all of the flashes of micro-lightning which will short out the circuit due to its excessive nodal voltages.

Nathan B. Stubblefield

Full-size Ammann article.

Self-shorting will occur if the entire circuit (along with all of its bare connections, ie: junctions) are not sealed with insulation, and this insulation is not surrounded by a common grounding plate or conductive film which is connected to Earth ground. Or as an alternative (in the following examples), most of its nodes are already shorted to each other to form a common ground. These interconnections distribute their elevated voltages among each other without actually reducing their elevated voltages. This alternative to grounding all nodes to Earth through a mutual insulator can accelerate the buildup of nodal voltage and accelerate the reactive expansion of the duration of *apparent energy storage* within a circuit's reactive components.

Tesla Loved to Brag!

Tesla bragged about this: that he had created a generator that would last for five thousand years, have no moving parts, and had no prime mover! That's uncanny! *Yet, probably true.*

Well, ...

I suspect that, given the historical evidence, ...

1. Tesla was ignored prior to World War II and considered to be a crackpot by our National Government. Unless he gave speeches, demonstrations, showed up at dinner parties, or wrote articles for publication in public media, he didn't get any recognition at all. Tesla was a nobody! It was only during WWII that our Federal Government took notice of Tesla due to the military intelligence reports coming back from the war effort which indicated that the Germans had the best of Tesla's technology making Tesla's existence hazardous to his health, namely: he could be abducted by anyone or worse, he could be assassinated because he knew too much.

2. This means that, once our United States government realized what a prized personage they had in their midst, all of Tesla's patents had to be immediately scrutinized for National Security and sequestered under claims of confidentiality or prior assignment to military entitlement the instant Tesla died.

This is 20/20 hindsight, to be sure, but conforms with the facts.

Prior to WWII, we had two prominent inventors who did some miraculous things with their inventions. I'm thinking of: the Ammann brothers in 1921, and Nathan Stubblefield in the 1890s.

I usually like to focus on the Ammann brothers since they got far more publicity than did Nathan Stubblefield. Yet, Nathan may have replicated[8],[9] the device which Tesla bragged about by reading Tesla's patents. And all of this occurred in the 1890s approximately half a century prior to WWII and long before our Government hushed up some of Tesla's more fantastic patents.

It is said of Nathan that, although he was a poor and uneducated melon farmer from Kentucky, he had a veracious appetite for self-education. He read every electrical engineering book he could get his hands on. So, he was not under-informed. Not by a long-shot!

Well, ...

I have placed before you (http://vinyasi.info/mhoslaw/Parametric%20Transformers/2022/Dec/simplest-overunity-circuit-you-will-ever-see_v8b.cir) a simulation (https://ufile.io/73h83rl8) of what I believe is a replication of Nathan's handiwork and a replication of Tesla's boastful admission.[10]

It gives a big surge within one and a half to two and a half days of its start-up and then gradually loses its power following the dictates of thermodynamic entropy.

But what a surge! And the best part, is that this surge will be sent into the Earth. Most of the nodes of this circuit are buried underground. Only a few, node numbers 2 and 3, surrounding the secondary coil, are exposed above ground. Also, the Living Tree which is acting as an aerial which receives a mixture of high, medium and low frequencies which naturally occur in our environment (along with the unnatural manmade frequencies) transfers these frequencies into its roots situated nearby the Primary and Secondary coils of the Transformer of this funky power supply. And the nodal voltages of this two and half day eventful surge do not affect the Living Tree a whole lot. The tree is spared getting fried to death. This tree is acting as a sine wave generator, of sorts, in which *any* frequency is the right frequency to catalyze a surge.

And the best part is this ...

I ran a simulation of this over a span of 10,000 virtual years of simulator run-time! It's power level dropped by a mere factor of one million to one as it

Schematic (https://ufile.io/73h83rl8) plus analysis settings (http://vinyasi.info/mhoslaw/Parametric%20Transformers/2022/Dec/simplest-overunity-circuit-you-will-ever-see_v8b.cir)

diminished towards zero amplitude of wattage on the Primary and Secondary coils. Considering that the initial wattage was just under a whopping 5e+160, I'd say it has a long ways to go before it dissipates all of its power, wouldn't you!?

And, since this device only surges once, if ever its operator had to "reboot" its powerful surge, all that would have to be done is to disconnect and reconnect the Living Tree from the Primary and Secondary coils, or go plant another generator underneath the roots of an old oak tree and wait a year before using it (as was customary for Nathan).

10k years of output!

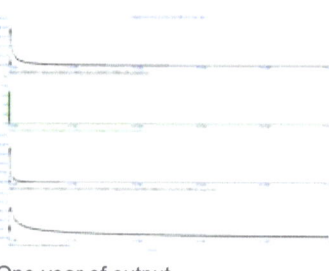
One year of output.

Background

Free Energy does not exist except within the mind of the beholder any more than does the concept of electrical engineering. Both do not exist within physicality. Instead, they exist as a set of imaginary constructs just like the concept of the square root of negative one is a made-up fiction born of the mind of man. Yet, these concepts have withstood the test of over a century of time and are assumed to be valid as a set of testimonials arising from the work-experience of countless electrical engineers despite our inability to physically prove any of it.

We also have some holes in our awareness[11] [12] about how physics addresses electrical engineering. And we also have some gaps of awareness within the domain of electrical engineering, itself, which assists with reemphasizing these broad-based holes along with the misconceptions which we create to make up the difference (by twisting half-truths into full mistruths). Please see the PDF file (to the right), entitled: *Breaking Ohm's Law to Achieve Overunity*.

BTW, I won't receive a Nobel Prize for my effort until well after I am dead since the recognition of my accomplishments any sooner than this would spawn a conflict between progress and keeping things just the way they are, which is: a general state of pervasive ignorance.

Almost Breaking Ohm's Law

BTW, a pseudoscientific topic is the refusal of science to investigate that topic for whatever reason or excuse. These excuses are not relevant to the refusal, itself. It's kind of like an act of excommunication by the Catholic Church emanating from a bygone era within "The Dark Ages" after the Fall of the Roman Empire and the fall of that culture which that empire had spawned. So, pseudoscience does not guarantee invalidity. Nor does it guarantee non-relevance. What it guarantees is pandemic ignorance along with misinformation to ensure that our collective ignorance and misunderstanding will continue to persist.

Premise

The argument surrounding "Free Energy" centers around a single Pro and a single Con. They are …

1. Pro – Answering the question: "Where does free energy originate if not from a mixture of a scarcity of externalized kinetic sources plus the reuse of reactive potentials?,"[13] and …
2. Con – How does "free energy" not violate the Conservation of Energy?[14]

Explaining the Con is easy since Conservation of Energy is a generalization of an ambiguous rendition of Kirchhoff's Current Law which is not a law so much as it is a vague generalization conveniently overlooking the magnetic coupling among individual inductors.[11]

By the way, Kirchhoff's Current Law does not conveniently forget (for the purposes of simplification) to include magnetic coupling to work in combination with nodal analysis of current passing through a node (to dissipate a pair of voltage differences) when formulating its so-called law. It purposely dropped any consideration of magnetic coupling due to the possibility that, sometimes, magnetic coupling may not be a conservable quantity.

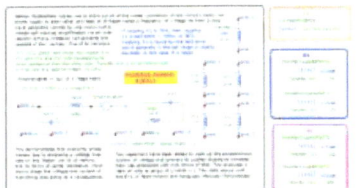

Most of the nodes of this schematic (https://ufile.io/5tc2xv8w) have been shorted to each other. This highlights how important is magnetic coupling within some types of overunity circuits.

Take this example (https://ufile.io/5tc2xv8w), for instance ...

Many of its nodes have been shorted out to reduce their relevance per Kirchhoff Current and Voltage Laws. The mathematical relationships among its various magnetic couplings are its significant feature.

These screenshots, and their Micro-Cap 12 simulation files, are located here ...

- Index of /mhoslaw/Parametric Transformers/2022/Nov (http://vinyasi.info/mhoslaw/Parametric%20Transformers/2022/Nov/?C=M;O=D) (vinyasi.info)

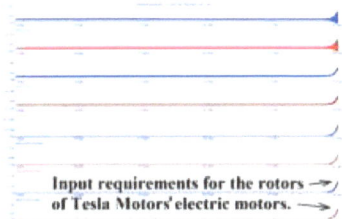

Virtual output tracings of a simulated circuit whose schematic possesses several shorts among its coils.

This leads to discovering where Free Energy arises from: the mutual inductances among a set of self-inductances – the very place where we have avoided looking!

This is what makes Paul Falstad's[15] simulator so unique by comparison to all of the other simulators available who are – for the most part – predicated on the Berkeley SPICE model.[16]

Paul's simulator possesses a different method of modeling transformers than does the Berkeley SPICE model making it very easy, *almost too easy*, to manifest free energy (http://vinyasi.info/ne) in circuitry. This enhancement should be given its proper place by assuming that his simulator is representing a very unique set of physical circumstances which are, as yet, unknown to most people including myself and Paul!

The predecessor to this realization was asking a series of questions all of which were related to the tolerance of solar panels to having high levels of current pass through them beyond their intended usage, such as: these two questions regarding the tolerance of solar panels to high current:[17] and[18].

A Rage Against the Storm

is.gd/ragingstorm

> **What are smart ways of reducing the amount of fossil fuels we burn, by stopping the supply, or by stopped the demand?**
>
> The trick is to eliminate the prohibition against the study of free voltage since free energy can readily arise from voltage by supplying somewhere for voltage to drain to, such as: a ground or a self-short.
>
> But deregulating nuclear power will produce more plutonium for more nuclear warheads to defend, what? Our prohibition against free voltage? Yeah, obviously we're maintaining the problem if we go the route which Abdel suggests.[19]
>
> Voltage can be had from anywhere in space by employing the use of an open transmission line concept/approach for its extraction. And reactance can accelerate the duration it will take to accumulate enough potential to run an EV. And a shorted transmission line concept will convert that accumulation of voltage into current giving us watts. Voila! Free energy from freely available voltage derived from empty space making use of severe reactance, ie. the reversal of current, to make it practical.
>
> Why aren't we studying this? 'Cuz it ain't good for the economy!
>
> Yet, it's good for us.

> A famous man once said, "Man was not made for the Sabbath; the Sabbath was made for man."
>
> Likewise, ...
>
> We are not slaves to money. For if we are, then our goose is cooked!

Hence, ...

Free Energy *does not* Exist; but Free Voltage *does* Exist! [20] [21] [22] [23] [24] [25]

Listen to some Music while you Read

Would you like to listen to some music while you read? [26] [27] [28] [29] [30] [31] right-click to open in a new window

This is inspired by a need to soften the blows of sharpened intellect... [32] [33]

Disclaimer

This wiki-book is not intended for anyone who lacks any background in basic electrical engineering (Learn more at Khan Academy (https://www.khanacademy.org/science/electrical-engineering/introduction-to-ee)), for it requires a familiarity with: Ohm's Law, Electrical Reactance, Complex Numbers, and their Polynomial Multiplication, basic electricity theory (https://pressbooks.bccampus.ca/basicelectricity/), and familiarity with circuit analysis (https://www.khanacademy.org/science/electrical-engineering/ee-circuit-analysis-topic) performed by electronic simulators. Without these skills, you'll be lost trying to understand whatever I have to say. You'll be perplexed anyway *even with these skills* since nothing you learned in school will have adequately prepared you for what is about to unfold...

There is no guarantee you will understand any of this. So, read through it -casually- once in a while without trying to grasp my intentions. Repetition, with breaks in between, might help.

Here's an example; also, a trick question. See if you can answer it?

Colin Mitchell's answer to: How many watts is a Farad? How many watts is a Henry? (https://www.quora.com/How-many-watts-is-a-Farad-How-many-watts-is-a-Henry/answer/Colin-Mitchell-35?ch=10&oid=397836189&share=9fa57c2e&srid=3zXxz&target_type=answer) on Quora.

Ohms law wheel WVOA

Definition

Free energy is a colloquialism suggesting getting more resultant energy exiting a device per energy expenditure which powers it. Yet, the mathematical concepts which promote and maintain our *rebellious belief in "Free Energy" do not exist* and neither do the mathematical constructs of electrical reactance. Both are fictions whose theorized existence have weathered our doubts for over a century of experience among electrical engineers encompassing a belief in the practicality of imaginary numbers.

The testimonials of numerous scientists and engineers (who attest to the practicality of their use of imaginary, and complex, enumerations within their calculations) does not prove the existence of imaginary numbers, nor does it prove that they succeed at representing any variety of electrical reactance, free energy or otherwise. And no testimonial has been put forward (by anyone) that imaginary numbers are useless. On the contrary, they are very useful and satisfy the need for using them. This demonstrates that we can "get by" without having to prove how to take the square root of a negative number. No one has a clue how to do that, and nobody expects to find out any time soon...!

Testimonials and demonstrations are no substitute for a well-constructed proof; and neither are arguments.[34] Testimonials are merely opinions, demonstrations are mere shadows of an understanding, and arguments are an attempt to promote a concept and all three are outside the jurisdiction of provability.

A proof demands an understanding which we fail to possess concerning the existence of imaginary numbers. And rationalizations for their usefulness does not substitute for lack of any proof.

Yet, so long as imaginary numbers serve us as a useful tool to temporarily hold an unprovable value, we can continue to use them so long as we never entirely forget that we are assuming the existence of a fantasy for the purposes of practicality.

Without concrete proof for the existence of imaginary numbers (in the world of physicality to which we are born), we will continue to have no physical proof for the existence of free energy, and no physical proof for the existence of electrical reactance since the two are closely related. (By the way, Free Energy is a special case of the more generalized topic of electrical reactance.) All we know is that the math works out based on over a century of "street-wise" expertise.

But the situation gets worse…

Free energy, if it is defined as a special case of electrical reactance, is a fantasy lacking testimonials since we also lack an understanding. The intention of this wiki book is to: *stop assuming that free energy does not exist and begin to seek an understanding by talking about it in rational terms which parallel our discussions of electrical reactance.*

Acknowledgments

The only reason why two paragraphs, up above, are so harsh-sounding is due to the peer-pressure, under which I have been operating, exerted by various editors over at Wikibooks who have managed to pressure me into deleting that text for various reasons, one of which is: *its lack of "relevant sources" (citations) to back up my theories.*

Well, …

I always assumed that I don't need to cite anyone other than myself – not due to any presumed brilliance of mine, but – due to common sense and simple logic.

The problem is, that I have a short-term memory in which (at my age) I don't always remember everything which I learn from other people – especially if I don't assign any significance to those new ideas at the time that I first hear or read about them.

Not to fear!

The memory of something which Aaron Murakami (https://aaronmurakami.com/) has said (on more than one occasion) has come to my rescue!

During one of his Energy, Science, and Technology Conferences (https://energyscienceconference.com/) given in Idaho, I was a witness to his use of the following analogy …

> Suppose that you spend one unit of energy per unit of time, and if you alter the unit of time in which that unit of energy is spent (without altering that quantity of energy per unit, nor alter its number of units), then doesn't this stand to reason that you've altered the rate at which you spend energy and, thus, altered the total quantity of its expenditure per its original unit of time?

For example, …

If I perform a single caloric unit of energy every time I strike my hand with my other fist, and I am doing this per second, then if I should increase my frequency of strikes per second, doesn't this increase the quantity of energy which is delivered to my hand (by my fist) during each period of one second?

This is why we're charged for our electrical energy usage using Ohm's Law – in Kilowatts – blended with per units of Hours since time is one of the three variables of Electrical Reactance Formulae. Time matters to electrical engineering and is not to be undersold since it works in conjunction with the other two ingredients of electrical reactance, namely: capacitive and inductive reactance.

Electrical reactance formulae don't bother to measure themselves using units of amps or volts. And, likewise, Ohm's Law doesn't bother to quantify itself in terms of the temporal-factors of frequency and phase relations, nor with the dynamic field-properties of capacitance and inductance. Yet, it is the combination of these two mathematical relationships which quantifies our electric bill.

Think about it!

If fact, even though we don't see capacitance or inductance on our electric bill doesn't mean that these parameters are not there by way of their implication. These parameters are assumed, according to certain criteria, by the electric company (who provides our electricity) based on a century of expertise of what to expect of its residential and industrial customers and how the electrical utility grid will *reactively* respond to those customers (considering the capacitive and inductive reactances which result from various types of consumer and industrial loads imposed upon the utility grid). And these presumptions generally hold true with minor deviations from one location to another and over time.

Now, ...

Parametric amplification[35] alters energy usage by modifying any one or more of the three parameters of electrical reactance, namely: capacitance, inductance, and frequency. Phase-shifting the time-component of when a wave of voltage or a wave of current peaks and troughs (nadirs) is also included within the time component of electrical reactance since it modifies the temporal relationships among the frequencies of voltage and current.

Parametric amplification manages to alter energy usage due to its ability to modify the capacitive and inductive fields which surround electrical components. And it alters the frequency at which oscillations occur causing these changes of energy usage to occur at faster or slower rates. This is significant since, if the rate of parametric amplification can supersede thermodynamic loss per unit of time, then overunity can be achieved.

It matters what the capacitance and the inductance of physical components are. But it also matters how one capacitor can modify the capacitive field of another capacitor and vice versa, and likewise for inductors modifying each other's inductive fields. They are able to do this because they are modifying their mutual capacitances and their mutual inductances which are just as important as their self-capacitances and their self-inductances.

Well, ...

Assuming that you accept the analogy, up-above, of energy expenditure per unit of time along with its consequences, and since time is merely one of three variable parameters of electrical reactance, then it stands to reason that *free energy* is the manipulation of <u>*all of the parameters of electrical reactance*</u>, not merely one of them (ie, time), while diminishing the significance of energy (under Ohm's Law) even though *some energy, no matter how small and insignificant*, is <u>always needed</u> to **run a circuit**.

In order to accomplish this feat of electrical engineering, two criteria must be met (which are repeated and explained further, below) in which ...

1. The input of energy must be kept extremely small (to diminish the significance of energy), and ...
2. No terminal of output should be allowed. This constitutes a *half-portal network*, or a *single* (one) *-terminal network,* (if such a concept exists; if not, then here is a new concept ;-). A more conventional *multi-terminal network* would encourage the formation of current along with entropy (thermodynamic losses). But a more constrictive *single-terminal network* (which is also starved of adequate input sufficient to power its loads) *may* encourage the formation of the reversal of current whose consequence is the magnification of voltage differences driving a circuit into becoming its own power source. This transforms passive components (particularly: inductors) into active components not requiring any significant Prime Mover (https://www.merriam-webster.com/dictionary/prime%20mover) other than a Prime Mover acting as a mere stimulant which catalyzes parametric amplification or parametric diminishment irrespective of thermodynamics (which is a separate consideration apart from parametric alteration of a circuit's energy).

You never heard of the expression, "Energy IN must equal energy OUT per Unit of Time", have you? Why not? Because it's expedient to understate the jurisdiction of the law, in physics, in which Energy (and Charge) must be Conserved. This misrepresents the significance of Kirchhoff's Voltage Law[36] and Kirchhoff's Current Law[37] as if electrical reactance never occurs.

<u>*Electrical reactance always occurs*</u>, to one extent or another, within every circuit no matter how mundane. Even a simple flashlight circuit exhibits electrical reactance by demonstrating inductance (and, thus, inductive reactance) along its length and capacitance (capacitive reactance) across its insulated boundary against its surroundings.

For what agenda are we, thus, brainwashed into avoiding a robust viewpoint? The answer is: the manipulation of society through the channels of government, commerce, education and entertainment, etc, for profit and self-glorification through ignorance of anything greater than established opinion.

We are being sold **half a bill of sale** whenever we hear the term of: *Conservation of Energy*. Yet, we're paying full price for this loss of *the other half* of electrical reality within the realm of electrical reactance.

Synopsis

is.gd/OUcriteria   **OR**   is.gd/oucriteria

There is a conspiracy taking place among theoretical scientists suppressing the virtual reality of free energy simulations by awarding those simulations a stigma of foolishness and foppishness while the theoretical scientists adopt an irreverent attitude that *free lunches are not worth studying and their ideologies are not worth promoting.*[38]

Well, in the physical world of consumerism, there are discounts all the time. Shoppers love them!

> Buy two; get one free!
> Half-off sale!
> Etc.

These promotional sales may not be an opportunity to walk out of the store with free merchandise, but it's definitely better than paying full price!

This *conspiracy* (derived from our collective ignorance and misrepresentation of Free Energy) carries over into our collective sensibilities as if the virtual world of electronic simulation cannot be taken as a guide on how to extricate ourselves from prevailing opinion.

Standard *physical theory* concerns itself with electrical engineering. Its presumption is that you have to, I repeat: HAVE TO, calculate the demand which a load will make upon a supply, and -then- add up all losses due to inefficiencies. This total must be, I repeat: MUST BE, supplied by the power source unless you want your physical appliance to fail.

That's nice. Yet, it merely describes the REAL POWER side of the physical *energy equation* as if ELECTRICAL REACTANCE was not a *virtual reality* worthy of our attention. The responsible practice of scientific inquiry attends to the details while never losing sight of the *big picture*.

It turns out that **reactance is** extremely, I repeat: EXTREMELY, **shy**. So much, so, that it doesn't take much voltage supplied by a virtual power source to suppress an *imaginary* reactance and prevent the eruption of unlimited oodles of freely available reactive power (fabricated from the square roots of negative one) which, whenever passed through a resistive load: such as a heater element, converts *invisible reactance* into REAL POWER miraculously convincing us that free energy exists when (in reality) free energy does not exist all by itself.

Free energy is a composition, over time, of the non-suppression of electrical reactance immediately followed by its conversion into usable power.

That's the conspiracy intended to keep all of us ignorant of our virtual options.

This "free-energy option" involves our admission of the virtual existence of electric and magnetic reactance (predicated upon the virtual reality of imaginary numbers) making it look as if (the conversion of reactance into) energy miraculously appeared out of nowhere when -instead- (what happens, is that) reactance (being lossless) cannot be spent nor lost. It must, thus, accumulate unless converted into a usable format (ie, energy). The accumulation of lossless reactance makes reactance the easiest, most available form of renewable energy.

Yet, feeding a virtual circuit too much (ie, conventional expectations of) voltage when that circuit is especially designed to take advantage of this free form of proto-energy (ie, reactance) will guarantee its failure to convince anyone of what I am saying is true.

Also, encouraging a throughput of current (through a conventional circuit), giving it an exit for current to pass out of, will guarantee suppression of free energy. This is in contradistinction to the restriction of the terminals of entry and exit to **merely one terminal** which is *exclusively utilized* as an inlet for the circuit's source of voltage and **this same terminal is** utilized as *the exclusive* outlet for its resulting current.

So, ...
Two criteria will guarantee the suppression of free energy under simulation ...

1. Feeding a simulated circuit too much voltage, and ...
2. Allowing the application of voltage to result in the flow of current by providing an exit.

Avoiding bullet points #1 and #2 will not guarantee the simulation of free energy since you also have to know how to take advantage of their avoidance whenever designing a virtual circuit. *But adhering to both points will guarantee the circuit's suppression* of free energy.

> WARNING — These criteria are intended to garner success *under simulation* and usually within the context of the Berkeley SPICE family of simulators[16] (but not all the time; other simulators[15] are, also, useful depending upon the situation). Although they are supported by standard mathematical criteria describing the conventional

engineering of electrodynamic theory, they are not intended to qualify the physics[39] behind these simulated strategies: That implication is left to the reader to vindicate, or not, through verifiable experience at your own risk of safety and success. *User, beware.*

Time Stands Alone: Space cannot Exist without Time.

Electric and magnetic reactance exists within the domain of time apart from space.[40]

Electricity exists within the domain of time and space.

Space is where Conservation of Energy occurs. Without space, conservation cannot be qualified nor can it be quantified. In fact, the opposite occurs wherein reactance must become altered over time when space is not involved, because energy does not exist outside of space.

So, when energy withdraws itself from space, all that remains is reactance. Thus, reactance exists -all along- coexistent with energy when both exist in space. But withdraw space from any consideration, and energy fails to justify itself without a spatial framework to give it a definition.

Within time, outside of space, reactance continues to exhibit the properties of inductance and capacitance. We would normally associate inductance and capacitance with the spatial phenomena of coils and capacitors which spawns them. But this is due to the inherent property of reactance which exclusively persists within the field of imaginary numbers and whose purveyance is the field of oscillatory time (as measured by the 2π angular momentum of each cycle of oscillation). Thus, inductance and capacitance *are never required to be real physical properties* despite the physical causes which we associate with them. Inductance and capacitance are non-physical properties of how time affects these properties and without any regard to space since these properties are not energetic properties; they affect energy without being energy, themselves.

Time has that impact upon spatial considerations: it affects spatial considerations without any allegiance to space since time dominates space.

The angular momentum of 2π binds the frequency of electrical reactance to time by defining each cycle of oscillation.

Inductance and capacitance do not require space to maintain themselves. The oscillations of time remembers them by converting their reactive output (resulting from prior cycles of oscillation) into the inductance and capacitance of subsequent cycles of oscillation. If reactance were somehow retained within the field of space, then this feedback could not occur. Space would, thus, conserve inductance and capacitance from one cycle of oscillation to the next. And this type of electric and magnetic reactance would be complex, rather than imaginary, since inductors and capacitors would be storing this reactance. But = *in the alternative* = the imaginary portion of electric and magnetic reactance can stand apart from space if the influence of real power is insignificant as to be of nearly zero amplitude. Under these ideal conditions, reactance feeds on itself creating more reactance from less reactance or, in the alternative, shrinks preexistent volumes of reactance (as the case may be) never reaching infinity, nor reaching zero, amplitudes of reactance due to this tendency for reactive feedback to become a multiplicative, or divisional, *trend* whenever real power is an insignificant input of apparent power.

As an aside :::

> We spend direct current during one-half of an oscillation and we recharge, or replace a spent charge with a fresh new charge, during each alternate cycle of oscillation. Thus, Direct Current is a subset of Alternating Current in which we casually, and conveniently, ignore the recharge, or replacement, phase of each cycle of Direct Current paying exclusive attention to each half-cycle of Direct Current which **spends** energy! But this is a game of make-believe in which we hide ourselves from the whole truth. Never, once, do we bother to seek it. Maybe this is why we encourage a way-of-life in which we throw away energy after using it merely once!? Ugh :::

Space is an extension of time which manifests electrodynamic phenomena in order to derive space from time.

Time can withdraw itself from space. When this happens, electricity vanishes leaving reactance in its wake.

Likewise, time can extend itself into space. When this happens, electricity manifests out of nowhere since time does not exist as a property of space. Nor is time a consequence of space. Quite the contrary! Space is a consequence of the electrodynamic extension of time.

In other words, space exists in time and coexists with time. But time is sufficient unto itself. This is where reactance occurs: in time, whether or not space is participating (and cooperating ;-).

But space must participate with time if electricity is to manifest itself. And conservation must participate (as well) within a framework of space cooperating with time in order for electrical energy to materialize.

Since energy has its equivalency within matter, one cannot exist without the other. Both energy and matter coexist, simultaneously, as variations of space. In fact, matter can never be lacking of an energetic state anymore than energy could lack matter to materialize energy since both are qualities of space. Hence, massless photons do not exist. Please see, the Appendix: Photons do not Exist.

So, ...

If you want to create energy, or create matter along with its dynamic aspect of energy (energized matter), you don't create matter (or energy) from space. Instead, you create a new space within a preexisting space for new matter (or energy) to exist within by extending both energetic matter – and its containment within space – from time.

Time coexists within all of matter and of energy. So, time is the ultimate source for the creative process to occur. And this temporal condition possesses the quality of electrical reactance from which electrical energy and the physicality of matter arises within their containment of space.

If We Can't Understand *Energy*, Then How Can We Possibly Understand *Free Energy*!

Introduction

The non-existence of Free Energy is not a lie so much as it does not also state that most of electrical engineering dabbles in non-existential reactive power predicated upon imaginary numbers which were invented by Hero of Alexandria to solve intractable problems and avoid the liability of proving their existence in the physical world.[41] In other words, what is the *physical* manifestation of the solution to... $i = \sqrt{-1}$ is a question which has yet to be answered by anyone.

Imaginary answers are not provable since they cannot be measured with physical instruments. They can merely be inferred by the mathematics of complex numbers as possibly existing somewhere in a fictional world often called, "counter-space" wherein everything is backwards (similar to Lewis Carroll's, "Alice in Wonderland", and, "Through the Looking Glass") in which elongated distances between the plates of a capacitor in our world of *space* is shrunken distances in counter-space.[42]

Fig. 0 – Triangular waves do not saturate inverted current (relative to voltage). They *must* escalate the production of negative wattage.

Free energy is not energy, yet it is freely available as a special case of reactive power, namely: an extremely low input of real power (nano watts or pico watts) fed into a circuit which lacks a throughput, see: Fig. 1a. This results in the reversal of current traveling backwards towards higher potentials of voltage resulting in the accumulation of a greater difference between those greater potentials and lesser potentials nearby, see: Fig. 0.

Block Diagram

is.gd/blockdiagram

Consider a circuit whose source voltage has merely one of its terminals connected to a circuit (constituting its input) while the other terminal (of this source of voltage) is connected to ground and there is no other ground connected to this style of circuit design (for the purposes of this hypothetical discussion, please see: Fig. 1a on the left and compare it with Tesla's design, Fig. 1b, on the right).

This configuration (of the terminal connections of a source of voltage feeding a circuit) discourages the manifestation of current which normally flows *into* a circuit through one terminal and flows *out* through another terminal. Instead, a restriction of terminals to merely ONE (in addition to severely restricting the input power) encourages breathing without flow, namely: the circuit manifests a standing wave in which the voltage and the current are out of phase by one-half cycle of oscillations. In other words, whenever the peak of voltage bounces off of the periphery of this type of circuit, the peak of current is crossing its imaginary center. During the subsequent half-cycle, the inverse occurs in which the peak of current echoes off of the periphery

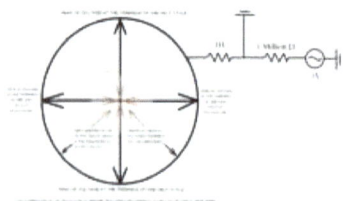

Fig. 1a – Oscillations of Radiant Energy due to throwing away most of the input and prohibiting the formation of current (within this circuit) by disallowing an exit (to avoid satisfying its inlet). For neophyte designers of overunity circuits, there should be only one inlet doubling as its own outlet.

at the same moment that the peak of voltage crosses the center. This creates an expansion, followed by a contraction, but not in the real world of physicality since the incentive for expansion (voltage potential) and the execution of same (its movement which reflects a flow of current) occur at opposing halves of each cycle of breath (so to speak)!

All of this occurs within the complex field surrounding reactive components.

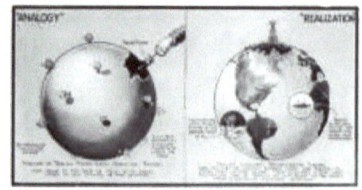

Fig. 1b – Tesla wireless power theory - Electrical Experimenter Feb 1919.

Although a circular pathway is avoided that would lead from a "source" to a "load" and then, back to the same "source", circular pathways are encouraged within the body of this style of circuitry so long as the various sub-circuits are electrically isolated from each other with merely a mutual inductance between them, and/or a single wire of electrical connection without any return path. These electrically isolated, open pathway, sub-circuits perform very well if they interconnect via several mutual inductances to make up for their lack of electrical connectivity.

This situation is best described as when **an open IDEAL**[43] **transmission line is terminated by one (or more) shorted IDEAL transmission line/s** encouraging the formation of a *purely imaginary impedance at the input*.[44] [45] [46] [23]

A simple four-arm cross symbolizing four open transmission lines emanating from out of a commonly located, central node.

Yet, this ideal condition[43] is not a fantasy. A bygone era of inventors[8] [9] [47] utilized magnetic remanence to preserve the magnetic field (which surrounds current) by incorporating the use of ferromagnetic materials wrapped around bare copper cable (placed directly underneath its insulation) over a hundred years ago (please see the figure of a Mu-metal cable, below-left) to prevent distortion/dispersion of the dots and dashes of the Morse code which was being sent across newly laid trans-Atlantic telegraph cable in the mid-1800s. We no longer use this method (probably) so as to avoid eddy currents and the inductive heating which ensues? Instead, we promote the use of copper or aluminum cable both of which lack the ferromagnetic preservation of current.

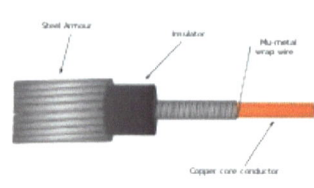

Mu-metal cable

There is no limit to how much mass of ferromagnetic material can be added to a circuit to preserve its current from thermodynamic losses since magnetic coupling can extend this illimitable mass to the area immediately adjacent to a circuit's coils.

An ankh-symbol is a cubic spline (https://towardsdatascience.com/cubic-splines-the-ultimate-regression-model-bd51a9cf396d) in which three of the straight-arms are symbolic of open transmission lines while the fourth top-most crossarm is representative of a shorted transmission line which loops back upon itself.

Bill Lyne quotes Nikola Tesla in his book, entitled: "Pentagon Aliens", as having said: ... *for every 200 pounds of iron which was magnetically coupled to* Tesla's Special Generator (https://www.gutenberg.org/cache/epub/39272/pg39272-images.html#Page_486), *one horsepower was increased at its output*.[48]

This style of circuit design (which I am espousing) tends to make it very easy to manifest an inversion of current 180 degrees out of phase with voltage. This inversion of current is oftentimes mistaken for its homologue of the "negation of resistance" which is mathematically equivalent, but not very educational.[49]

A more accurate description would be the *negation of reactive voltage divided by impedance*, namely ...

$$\text{Negative Current} = -\left(\frac{\text{Reactive Voltage}}{\text{Impedance}}\right)$$

This leads to another, more traditional, version of Ohm's Law in which Power equals Voltage Squared Divided by Resistance: $P = \dfrac{V^2}{R}$.

That conventional version is vague and incorrect in so far as it does not distinguish what is occurring, namely, that: Negative Watts is equal to the Application (the Input) of Real Voltage times its Resultant Output of Reactive Voltage divided by various Impedances (both Real and Imaginary) within a framework of time …

$$\text{Negative Watts/Unit of Time} = -\left(\frac{\text{Real Voltage Input} \times \text{Reactive Voltage Output}}{\text{Impedances}}\right) \bigg/ \text{Unit of Time}$$

Granted, this is a more convoluted restatement of Ohm's Law with the distinct advantage of sidestepping the conventional claim of physics in which: "Energy IN equals Energy OUT" by (instead) implying that: "Real Voltage IN *cannot equal* Reactive Voltage OUT". The resulting reaction of output voltage **must be** greater than, or less than, input voltage irrespective of thermodynamics. This is in contradistinction to conventional wisdom since (my perspective is that) the input is complex and the output is also complex *all the time* (a real value plus or minus an imaginary value). In fact, all circuits possess some reactance in proportion to some non-reactance. This is why I deem the traditional presentations of Ohm's Law flawed (in principle) while maintaining a more practical approach for technicians by avoiding a fundamental teaching of how electricity behaves.

Without this fundamental understanding, no one will appreciate Free Energy since they will lack a robust understanding of electricity. I challenge everyone, who desires an understanding of Free Energy, to return to basics and rethink what we've been taught. Ergo, current is a fiction. It is a mathematical shorthand notation replacing something slightly more complicated.

Fig. 1b suggests a similarity to Fig. 1a. Both images possess a singular inlet for power resulting in a periodic variation of potential occurring everywhere, simultaneously, and without any manifestation of conventional current (subject to entropy) that could delay and reduce (through losses) the transmission of power. On the other hand, the reversal of current (in this wikibook's proposal), produces the inversion of losses, namely: an escalation of gain.

This is similar to if, **whenever we shop at a market, <u>they pay us</u> to take their groceries instead of charging us! And… Every time we shop, <u>they pay us more than they paid us before</u> while claiming to pay us the same!** *{The inverse of deprivation.}* **What a trip! With so much abundance, who needs war?**

Mathematical Consequences

A Low Input Power

Lots of Real Power, plus or minus, a modest amount of reactance will guarantee the conventional stability (or, Rule of Thumb) that reactance cannot grow by way of feeding itself from the reactive field surrounding reactive components, such as: inductors and capacitors, resulting from the outcome of the prior cycle of oscillation since excessive real voltage will suppress a runaway self-looping of electrical reactance.

Yet…
Severely restricting the use of real power at the inlet of a circuit's source of energy will encourage the unconventional rule of thumb in which electrical reactance will be almost exclusively nourished by its own feedback irrespective of thermodynamics or the Conservation of Energy – especially since energy plays no significant role, here, since its input is severely limited to be less than a micro watt.

In other words, any complex number (enumerating the amplitude of either a wave of voltage or a wave of current) possesses two components: a real number and an imaginary number. The magnitude of the real number regulates the consequence of how the present cycle of oscillation impacts any subsequent cycle. Meanwhile, the imaginary number can create the inversion of current when squared if the self-looping, self-feeding tendency of electrical reactance is not suppressed by any excessive input of real power.

If this complex polynomial…

$$A \pm B \times \sqrt{-1}$$

…is squared…

$$(A \pm B \times \sqrt{-1})^2$$

…then, the result is four products reduced to three (since two results, the cross products of A times B, are similar enough to be grouped together) …

1. The square of the real component, A^2.

2. The cross-product of the real portion times the imaginary portion, $\pm 2AB\sqrt{-1}$.
3. And, the square of the imaginary component, $(B \times \sqrt{-1})^2 \equiv B^2 \times (\sqrt{-1})^2 \equiv B^2 \times -1 \equiv -B^2$.

If the real power input of A is restricted to a very small value, of nano watts or pico watts, then the negation of real power resulting from the squaring of the imaginary portion of this complex number will not be oppressively regulated out of existence. Only the tiny value of A will shrink or maintain its amplitude while the amplitude of $-B^2$ will grow at an exponential rate. By restricting the inlet of real power (feeding this style of circuitry), there will be an increased likelihood of success in producing radiant power serving as a precursor to free energy. Yet, this is not all that is required to ensure success.

It is also necessary to connect only one terminal of a voltage source to this type of circuit while connecting the other terminal (of the voltage source) to ground and disallow any other ground to be located anywhere else within this type of circuit (in the beginning if you are not yet "skilled in this artistry."). This will ensure that no current forms since it won't have anywhere to drain after scantily-leaking into the circuit from the voltage source. This will ensure a radial pattern of circumferential pattern of peaks and troughs, in which the peaks of voltage will bounce off of this circuit's periphery at the same time that the peaks of current will be crossing the virtual center of this type of circuit during each half-cycle with an inverse pattern at the next half-cycle.

In other words, current has been divested of its singular significance. Only voltage matters, plus: frequency, phase shifting of voltage versus current, plus inductance and capacitance.

Since the inherent tendency of electricity is to make up the difference for any shortcomings, current will form (anyway) despite our best efforts at preventing it. This "last ditch effort" on the part of "nature's tendency" will ensure a reversal of current since that is the only direction we will have allowed for *by failing to prevent it*.

Ponder this ...

If, after taking every precaution to prevent the flow of current, don't you think that the only other option available (to Mother Nature) is for current to flow backwards as if in rebellion to our various restrictions?

In the words of actor: Jeff Goldblum's character, portraying a mathematician who specializes in chaos theory in the movie, *Jurassic Park* (part one): "Life will always find a way to break free of any loss of liberty."

> Conservation of Energy *is a status symbol confessing* allegiance to the herd *since it is grounded in physical reality as constituting the ultimate and exclusive verification for any authority while simultaneously ignoring electrical reactance subsisting within the domain of time acting as the trump-card (so to speak) giving us the liberty to recycle energy rather than blindly throwing it away (returning it back to its source) after every single use and refusing to* pay through the nose *for this wasteful method of consuming energy*.

Whoever conjured-up this scheme must be a madman!

It sucks!

Voltage Drop

It stands to reason that electrical voltage drop is a mathematical process which cannot be performed upon the imaginary coefficient of a complex polynomial. It may only be performed upon its real number coefficient. This is a consequence of the assumption that voltage drop is the distribution of a real numbered evaluation of voltage across a circuit resulting from simple resistance rather than from electrical reactance.[50] This allows for the accumulation of reactive potential as well as for the accumulation of reactive impedances (both inductive and capacitive). This latter accumulation can occur within the imaginary fields surrounding reactive components only if the distribution of real voltage is kept below useful values amounting to nano watts and pico watts so as to avoid disturbing (suppressing) reactive feedback. This accumulation of reactance serves as feedback for the input of subsequent cycles of oscillation causing reactance to escalate at exponential values. Hence, "free energy" is an incorrect assessment of this peculiar situation. A more rational explanation is to claim "freely available reactance" resulting from an extremely low input of real power.

Convention teaches us that the peaks and troughs of voltage and current may oscillate their amplitudes as they travel around the circumference of a circuit. But there is another possibility in which they may echo their peaks and troughs in diametric opposition to each other during each half of an oscillation effectively creating a standing wave of one-half cycle of displacement between their phases (See, Fig. 1a, above). This will only occur if we discourage or prohibit the formation of current while maximizing the accumulation of the imaginary component of reactive power. At some point, the complex enumeration of the real and imaginary portions of electric power will be squared during our mathematical assessment of the electrodynamic behavior of a circuit. If we keep the input voltage extremely low and suppress the flow of current, then we may succeed at developing more reactance than what conventional wisdom would expect. And when, through simple (thermodynamic) conversion when passed through a resistor, the complex result (of the squaring of a complex value) will have its phases of real voltage realigned with its phases of reactive voltage and with its various impedances (voltage realigned with current possessing a power factor of positive unity, +1) and, thus, be able to convert the cross-product of: $\pm 2AB\sqrt{-1}$ into the squaring of the imaginary portion: $-B^2$, of a complex reactance.

Utilization of Electrical Reactance

What is the Ground Plane?

What is the Ground Plane?[51][52]

Up until now, I did not understand how a Berkeley SPICE[16] ground component operates. Nor did I understand its difference from how a ground operates in Paul Falstad's simulator.[15]

But the study of Byron Brubaker's version, 1.0, of Tesla's Hairpin Circuit[53] prompted me to finally understand these differences.

The ground components in both simulators are unique to each other due to how a referencing ground operates in each simulator.

Schematic drawing (https://ufile.io/zsbuwg98) of a simulation (http://tinyasi.info/mhoslaw/Parametric%20Transformers/2023/Jan/Byron%20Brubaker%20Tesla%20Hairpin.zip) of Byron Brubaker's rendition of Tesla's Hairpin circuit.

In Paul Falstad's simulator, a referencing ground is already provided by the software. So, the user does not have to insert one anywhere since that has already been accomplished. The only purpose for the user to insert a ground (in Paul Falstad's simulator) is to act as a source for electronic flow of current. This presupposes that the Earth is a battery of around one microvolt at ground-level whose amp-hour capacity is vast!

A referencing ground is not provided in the Berkeley SPICE family of simulators, to which Micro-Cap[54] and LTSPICE[55] are members. So, this is the only function which a ground represents in these simulators. If you want to represent anything else, such as what a user-inserted ground would represent in Paul Falstad's simulator, then you have to insert a micro volt battery in between a ground and your circuit. You will also have to add a low-level capacitor behind the battery. I have labeled these, *CapacitanceOfEarth1* and *CapacitanceOfEarth2*. Between this battery and the circuit, you may position your load. Mine, here, is a capacitive load, labeled: *CapLoad*. It is *not* a resistive load because that cannot sustain the transient (a momentary surge) which this simulation suddenly provokes and quickly dissipates to zero. So, a relatively large capacitive load could retain its charge and a series of switchings could clear these charges to zero by transferring these charges to the actual load that you wish to power and then restart this circuit from another cold start?

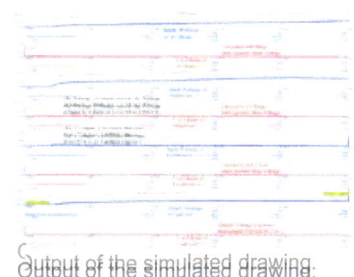

Output of the older, simulated version of Byron's Hairpin.

Raw schematic of an older version of a simulation of Byron Brubaker's rendition of Tesla's Hairpin circuit. Notice the complexity of this previous version versus the simplified drawing, up above? Don't build this version; build that one and safely play with modifying its parameters.

All of my capacitors possess 3 Ohms of equivalent series resistance. Both of my coils possess units of resistance equal to their units of inductance to approximate a wire gauge of 25 AWG.

The rate at which over-reactance exhibits overunity is determined by the smallness of the two capacitors, labeled: *CapacitanceOfEarth1* and *CapacitanceOfEarth2*. Although the Earth's capacitance is assumed to be ≈711 micro Farads, this does not produce overunity within a reasonable length of time. Or else, it may possibly fail to produce overunity at all. I don't know. I didn't have the patience to wait long enough to find out. So, to speed things up a bit, I chose to use values which are less than 100 nano Farads, such as: 99nF or 10pF or 100 femto Farads.

- BTW, one micro volt is the atmospheric voltage at ground level which is enough to power a crystal radio set from the 1920s.
- ACMains amperage is the amperage which is being drawn from the utility grid.
- MainsGnd is the amperage which enters the circuit from the ground adjacent to the ACMains.

- *EarthSource* is the amperage which the Earth is providing free of charge!
- *CapLoad* voltage dominates *CapLoad* amperage by a factor of 3 to 1 due to this component's equivalent series resistance is set to 3 Ohms.
- BTW, if you choose to build either of these simulations, you may want to (somehow) engineer the prevention of their tendency towards exploding, suddenly, with an excessive amount of power rather than allow this circuit to fry itself (or throw shrapnel in all directions).

Latest, updated, schematic drawing (https://ufile.io/il0hiucg) of a simulation (http://vinyasi.info/mhoslaw/Parametric%20Transformers/2023/Jan/Byron%20Brubaker%20Tesla%20Hairpin.zip) of Byron Brubaker's rendition of Tesla's Hairpin circuit.

< < < < You can get away with shorting out the four terminals of this step-down transformer and you'd almost get the same level of output, but you'd be wasting a slightly elevated input power. Patent US 600,457 A (https://patentimages.storage.googleapis.com/de/5d/14/a57ffad14ccd94/US600457.pdf) for Nathan Stubblefield's Earth Battery may be on the left if we replace "ACMains" with a live oak tree and replace "MainsGnd" with its roots in the Earth not including what is missing from his patent to its right.

If you don't have access to the Earth to serve as your "ground plane", in the alternative, you can simulate the Earth's ground plane by utilizing a large conductive surface, such as: another antenna. This is what the Ammann brothers did with their Atmospheric Generator: they used two spherical antennas to serve as their monopole antenna plus ground since their mobile power station of their batteryless electric car of 1921 needed a grounded antenna and they could not provide an Earthed ground plane for their car.

The red arrows point to their dipole antenna substituting for Brubaker's pair of upside-down Earth grounds.

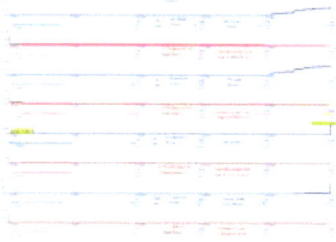

An output of 400k Tera Watts at the CapLoad versus an input of no greater than the EarthSource (contributing 23k Watts) is a ratio of output to input of 17.4 Tera to 1!

Output of the latest, simulated version of Byron's Hairpin.

In other words, turn this image (up, above) upside down and imagine its two electric grounds are two spherical antennas similar to what the Ammann brothers mounted on top of the drum-shaped power station which was strapped to the front-end of their car >>>>

A dipole antenna is similar enough to a grounded monopole antenna to serve as its replacement. An example of a dipole antenna is the spark-gap transmitter of Heinrich Hertz as it morphs into the monopole antenna of Guglielmo Marconi.

All four caps: C1, C2, C3 & CapLoad, possess 10 Ohms of ESR. Applying pressure will raise their ESR and is hinted in a Tesla patent for manufacturing capacitors under pressure, US Patent 577,671. (https://patentimages.storage.googleapis.com/a/bc/8c/ca152a00b76d05/US577671.pdf) Thanks go to Byron Brubaker for this tip! Surf to … Equivalent series resistance for an explanation of ESR.

Tesla also has a patent for using oil as a dielectric material for capacitors: Patent No. 567,818. (https://patentimages.storage.googleapis.com/b5/62/b7/bb55a0b2cf4518/US567818.pdf) Pressurizing this oil may be the easy way to increase the equivalent series resistance of a capacitor to alter its behavior in an overunity circuit? What do you think?

The output of virtual oscilloscope tracings to the right establishes an impedance which varies over time. This refutes any credibility that resonance is required. > > > > > > > > Rarely, does resonance have anything to do with overunity. For, impedance starts out very high at the beginning of all of the red tracings to the far-left in which voltage dominates over current. Yet, over the course of 32 milli seconds, that situation drastically changes in which current dominates over voltage due to a drop in impedance to the far right. The circuit actually alters its own impedance! Ask a tank circuit to do that! Yeah, right… This drop of impedance probably has something to do with the rise of wattage traced in all of the blue graphs?

WARNING — HIGH NODAL VOLTAGES WILL ACCUMULATE!

Capacitance of the Earth

The square root of the dielectric constant in a capacitor is a refractive index of that material. This suggests that capacitance is prismatic and that waves which interact with a capacitor are both refracted and reflected and, thus, spawn two daughter waves for every time that a parent wave interacts with the dielectric material within a capacitor. This is how complicated the behavior of capacitors are.

And, ...

This is why Eric Dollard, says that: *"The energy which enters a capacitor is not the same energy which exits it."* Eric has also said, that: *"A lightning bolt does not form from the energy which sits behind the voltage difference between its terminals, for that is not enough energy to fully charge up the lightning bolt. That's merely enough to form its ionic channel. Most of its energy enters into the lightning's numerous branches at right angles to each branch from the surrounding environment. This source of energy is literally "sucked" into each future branch by reactive forces acting as pumps for energy against gradients of impedance."*

I would venture a guess that the refracted daughter wave is imaginary (purely reactive) while the reflected daughter wave is (purely) real, and that an overunity circuit would ideally emphasize refractive capacitive daughter waves while conventionally thermodynamic, lossy circuits will emphasize reflective daughter waves since Micro-Cap help-file says, ...

> *"The ideal capacitor acts as a shock absorber and, thus, must not be too low in capacitance lest it fail to perform this task."* – In other words, an ideal capacitor must emphasize absorption and discharge of wave energy. Reflection is an indication of a combination of absorption followed by the discharge of energy."

Refraction of capacitive reactance is the ideal circumstance for us to encourage since it will magnify reactive power output so long as we minimize real power input. Then, somewhere along the way, the imaginary portion of the complex enumeration of reactive power will get squared. And when it does, it will leave a scanty quantity of real power, a build up of negative watts, and a remainder of complex power which will accumulate over time.

This build up of negative watts will not be figurative as in the case of the passive sign convention's definition of the generation of power. Instead, it will actually fulfill that definition in the physical world by making it possible for a passive inductor to become an active inductor and generate current without the necessity of engaging a prime mover's external source of energy in order to move that coil through a magnetic field and, thus, defy Faraday's Law of Induction by superseding it with a higher law whose text has yet to be written into our conventional wisdom.

The conventional presumption of the capacitance of the Earth is estimated to be in the vicinity of 711 micro Farads on the presumption that it is a solid sphere whose outer concentric shell is infinitely distant. But, what if it is hollow with a presumed thickness of 1 kilometer or less?

6.3675 Megameters = the average radius of the Earth

1 km = assumed thickness of the Earth's shell – probably less than this

"The refractive index of a nonmagnetic material is the square root of its relative permittivity."[56]

Since the dielectric constant, or: relative permittivity, of a material is the square of its refractive index, and the refractive index of granite bedrock is somewhere between 2 and 3.5,[57] then the average refractive index of granite is 2.75 and its square is 7.5625.[58]

Outer surface diameter of Earth's shell minus the thickness of this shell equals the inner surface diameter: $(6.3675 \times 10^6 \, meters) - (1 \times 10^3 \, meters) = 5.3675 \times 10^6 \, meters$

Capacitance of two concentric spheres in which the larger sphere has a radius of: $r1$ (representing the outer surface of the Earth) versus $r2$ (representing the inner surface of the Earth).

$$\frac{r1 \times r2 \times 4 \times Pi \times \text{the dielectric constant of a substance}}{r1 - r2} = capacitance^{[59]}$$

$$\left(\frac{(6.3675 \times 10^6 \, meters) \times (5.3675 \times 10^6 \, meters)}{(6.3675 \times 10^6 \, meters) - (5.3675 \times 10^6 \, meters)} = \frac{3.417755625 \times 10^{13} \, meters}{1 \times 10^3 \, meters} = (3.417755625 \times 10^{10} \, meters) \right) \times \ldots$$

$\hookrightarrow \ldots \times 4 \times Pi \times (7.5625 = \text{the dielectric constant of granite bedrock}) \approx 3.25$ Tera Farads ... or more ... !!!

Conclusion ...

Since my simulation (https://ufile.io/tid1tirs) requires the presumption of the maximum capacitance of the Earth is no greater than 99nF, this circuit may not give the expected results unless some modification can be devised which overrides the

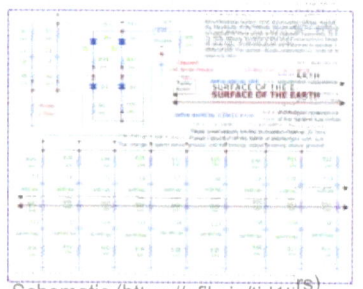

Schematic (https://ufile.io/tid1tirs) upholding Foster's reactance theorem for over-reactance driven by negative impedance.

Output of schematic. (http://vinyasi.info/mhoslaw/Parametric%20Transformers/2023/Jan/brubaker6h4.zip)

Earth's capacitance?

Maybe a 100 femto Farad (one-tenth of a pico Farad) capacitor should be built into each of the nine or more grounding rods above where each rod enters into the Earth? Yes! It works …

And, …Since this circuit tends to develop high nodal voltages …

For safety's sake, maybe a dielectric goo should be poured into this circuit's project box to solidify into an immersion of a dielectric medium after, first, lining the interior surface of the project box with conductive plates and each plate is separately (?) connected to its own Earth ground?

Where's the Energy Coming From?

::: within Byron Brubaker's Hairpin circuit (which he derived from Nikola Tesla's Hairpin)?

1. The full-bridge rectification via four diodes are generating it from within themselves! Excluding a ground on the left insures that this is the answer.
2. It's coming from the Earth. Placing a diode on each and every ground and inverting their terminals at the same time to create a switch verifies that this is the answer.
3. Take your pick! Whichever version suits your fancy since there's no way to prove which is the case. It could be either, or both.

If the Extra Energy is Coming From the Earth

The electric utility companies may be generating power within the capacitor bank, substations mentioned by Eric in a video excerpt extracted from a longer YouTube video, entitled: *Origin of Energy Synthesis*.[60] Although, he defines their usage is for bringing the phase relations of voltage versus the phase relations of current into a harmonious relationship so that transformers don't blow up and power is successfully delivered over hundreds of miles of transmission, in addition to his definition of their usage, the utility companies may have another usage of adding power to the grid along the way.

The video (within the footnote, above) is a companion to another related video (https://www.youtube.com/watch?v=4jACf j6sHK0&feature=youtu.be) which is on the topic of electro-mechanical watt-hour meters being used by the power companies to supplement energy coming from the power plant.

If you put the ideas of these two videos together, then it becomes possible to speculate that the power companies may have several techniques at their disposal to supplement power plant generation?

A critic assails me on YouTube, …

"Do you guys really think that little spinning disc is generating enough power to run an electric range, air conditioner, toaster, etc?"

My response, …

The catalyst need not consume much energy to function as a stimulus. Most of the value of a rotary generator at an electric power plant is not in the application of its voltage (which everybody misconceives is the case) but is embodied in the frequency of that generator. Because, if we double the frequency then the power will be quadrupled. So, think of the frequency of that little disc as being significant in that it replaces or supersedes the frequency of the line voltage that we are accustomed to being held at 60 cycles or 50 cycles as the case may be. But if we can get the frequency up to 1,000,000 Hz, then we're really creating a tremendous stimulus for the power to come out from somewhere else. Now, where may that power be coming from? It can come from numerous sources that we don't usually take into consideration, such as: capacitance, inductance, and even resistance! These are the things that I look at in a free energy circuit which I am designing under simulation because potential sources of energy are just as valid as are kinetic sources of energy (such as a prime mover). But we put too much emphasis on prime movers as being the sole source of the energy for a circuit when in fact a prime mover's significance can be reduced to that of a mere stimulus (a catalyst) provoking reactive power to come out of reactive sources such as capacitance and inductance and resistance. In other words, we are using impedance as our source of energy and we are using a prime mover as its catalyst:

Transmission Line Voltage

Transmission lines differ in how much voltage they carry based on their length, plus other factors:

1. Local transmission lines typically possess 13.8kV.[61]
2. Short transmission lines – In short transmission lines, the length is within 50km and the voltage is limited to less than 20 kV. In short transmission lines, the effect of line resistance and inductance is more predominant than capacitance.
3. Medium transmission lines – These lines have an overhead cable length of greater than 50km and less than 150km. The allowable voltage ranges from 20 to 100 kV. The analysis of medium transmission lines considers the three lumped line constants: resistance, inductance, and capacitance.
4. Long transmission lines – Overhead transmission lines with lengths greater than 150km and voltages above 100kV form long transmission lines. Line constants are considered distributed elements in the analysis of long transmission lines.[62]

Long transmission lines are greater than 150km and possess voltages greater than 100kV such as the exampled output (of the schematic on the left) exhibiting: 300kV on its indLoad (inductive load). The series resistance of this inductive load is: 6.24e+249 Ohms and its inductance is: 1e+250 Henrys, yet the V/I ratio of its output is: 7.5e+18! I'd expect something much greater, wouldn't you? Something more similar to its resistance of: 6.24e+249 Ohms!

Oh, well ... I still don't know everything!

Wasteful UFO Technology

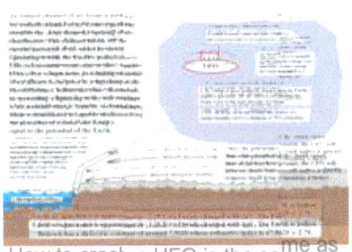

How to crash a UFO is the same as how to dispel a cloud using Wilhelm Reich's cloudbuster to accentuate the entrainment of ionized air.

If you're going to design an anti-gravity craft, you'd better make sure it runs on free energy since UFOs are very inefficient at their utilization of energy. They waste a tremendous quantity of energy to neutralize inertia, aka. gravity, since there are no motor coils from which to recover reactive power upon turning OFF the power supply of a UFO craft. Instead of coils, a UFO craft uses high voltage electrostatic fields surrounding their craft to substitute for their lack of coils. This is why Wilhelm Reich was able to crash a UFO by accidentally pointing his cloudbuster tubes at a cloud within which a UFO craft was hiding since the humidity was above 70% and the ions of air immediately surrounding that craft was of a positive polarity. These are the classic symptoms of a smoggy day in Los Angeles or San Bernadino Counties. Into this high voltage, electrostatic field is projected the electromagnetics which are necessary to neutralize inertia and gravity. But if this electrostatic field should disappear, then its UFO craft will come crashing to the ground! This is what got Wilhelm Reich arrested, but into a Federal prison, all of his books were burned in the United States, and he never made it out alive. He was murdered by an inmate just before his release date.

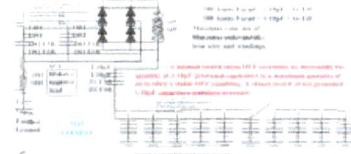

Generalized drawing (http://vinyasi.info/mhoslaw/Parametric%20Transformers/2023/Feb/brubaker8_manual-switch.zip) of a schematic for a free energy device which makes use of Foster's reactance theorem to over-react and deliver negative impedance. Real power is converted into purely imaginary, reactive power through the use of extremely low-level capacitors which makes their dielectric material partly prismatic owing to a material's refractive index is the square root of its dielectric constant.[56] Diodes are also used in this schematic.

Busting the Clouds of Ignorance (https://www.youtube.com/watch?v=kPYdeSZhmmY&feature=youtu.be) – a YouTube video of one-hour duration.[63]

How to Crash a UFO (https://www.youtube.com/watch?v=irpu9U2hhwc&feature=youtu.be) – a YouTube video of one-hour duration.[64]

UFO Power Supply

As stated above, a UFO power supply should *ideally* be of an overunity condition to justify the craft's wasteful expense of throwing away electric power in the form of the electrostatic force which is projected out into the immediate vicinity of the craft to safeguard against the loss of the electromagnetic force which is used for neutralizing inertia and gravity.[65]

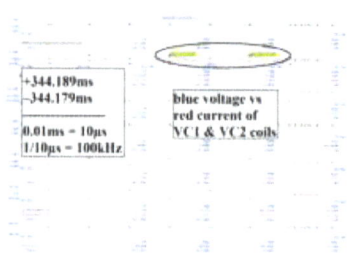

Closeup view and the math to calculate the frequency of the output on the main coils, VC1 and VC2.

It should, probably, also be pulsations of frequency since this is what was reported by a B-47 which was outfitted with electronic surveillance equipment during the 1957 UFO sighting written up in a book by James M. McCampbell[66] and discovered in a bookstore by David Alzofon and included in David's book[67] about his father's theory on gravity control[68] which is directly related to the topic of *Dynamic nuclear orientation,* by C. D. Jeffries. (https://www.bing.com/search?q=Dynamic+nuclear+orientation%3A+C.+D.+Jeffries&FORM=SSPRAC&PC=U531&lightschemeovr=1)[69]

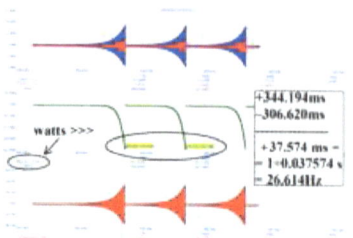

Duration of pulsations on the main coils, VC1 and VC2.

There were five criteria of this UFO sighting as sensed by the electronic equipment on board the post-war (WWII) renovated B-47 bomber. They were ...

1. Frequency: 2,995 to 3,000 Megacycles per second [2.9 to 3.0 Giga Hertz]
2. Pulse width: 2.0 microseconds
3. Pulse repetition frequency: 600 cycles per second
4. [rotational] Sweep rate: 4 rpm
5. Polarity: vertical

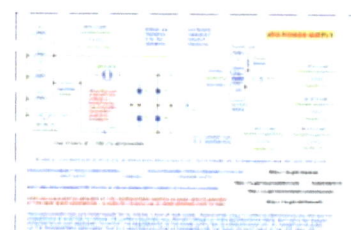

Conjectural power supply (http://vinyasi.info/mhoslaw/Parametric%20Transformers/2023/Feb/UFO_power-supply.zip) for a UFO craft.

Surprisingly, my simulation of two years ago, (http://vinyasi.info/mhoslaw/Parametric%20Transformers/2023/Feb/UFO_power-supply.zip) which I had included on a slightly unrelated topic on Quora,[70] is similar in some respects for being a pulsed frequency which sweeps its wattage upwards before repeating this behavior.

1. Input Frequency: 20kHz; 3V sine waves.
2. Output Frequency: 100kHz; triangle waves whose phases of voltage are displaced from its phases of current by one-half cycle of oscillation.
3. Output Sweeps its Increase of Amplitude per Pulse before collapsing back to the start of each pulsation.
4. Frequency of Pulsating Sweeps: 26.614Hz

Kate Bush's Tribute to Wilhelm Reich

- Kate Bush did an homage to Wilhelm (https://genius.com/Kate-bush-cloudbusting-lyrics) in the form of a music video, called: Cloudbusting, back in the 1980s with assistance from actor, Donald Sutherland. Watch it on YouTube: Kate Bush - Cloudbusting HD LPR remastering (The Whole Story 2015) (https://www.youtube.com/watch?v=AJIl-eF5zUA)

- Kate Bush - Cloudbusting - Official Music Video (https://www.youtube.com/watch?v=pllRW9wETzw) – original

- MTV interview (https://www.youtube.com/watch?v=0SQJrYfLztU)

- *A Book of Dreams:* The Book That Inspired Kate Bush's Hit Song: *Cloudbusting,* – June 1, 2019 (https://a.co/d/7ZCEwFl)

- Eighty-three photographs of my successful attempt to clear the sky of chemtrails plus the additional benefit of making it rain six days later! (http://w.earthinglife.info/media/index_02.html) I had the help of the empowering bliss of a yogi (https://www.tm.org/) along with Leon Ernest Eeman's biocircuit (http://vinyasi.info/texts/biocircuits/) and a P-gun derived from Peter Lindemann's design (length of tube versus its inner tubular diameter {bore diameter} is: the square root of two times 32 to 1) and covered in orgone layering (rosin paper coated with rosin alternating with aluminum foil and a convex mirror on the bottom end; the open end faces straight up; with an iron wire running down the length of its inside connected to two opposing points of a biocircuit {probably left wrist to throat} with me in it for 45 minutes) used by Trevor Constable. (https://t

revorconstable.com/) The other three pairs of biocircuit connections are: right wrist to crotch; right ear lobe to left ankle; left ear lobe to right ankle.

Links to a few Resources about UFOs

- If These US Navy Patents are Made Then We Are in a Star Trek Technology World (http://vinyasi.info/energy/If%20These%20US%20Navy%20Patents%20are%20Made%20Then%20We%20Are%20in%20a%20Star%20Trek%20Technology%20World.pdf) – shortcut URL ... is.gd/keviqu
- United States Patent No. US 10,144,532 B2; Dec. 4, 2018 (http://vinyasi.info/energy/US10144532.pdf) – on Google patents (https://patentimages.storage.googleapis.com/de/4c/43/62c585ccc936cc/US10144532.pdf)
- *Occult Ether Physics,* 4th revised and expanded edition: Tesla's Ideal Flying Machine and the Conspiracy to Conceal it, by William Lyne (https://a.co/d/0Rb5xsG)
- Organick Energy 7600 Channel on YouTube (https://www.youtube.com/@organickenergy7600)
- Paul Scarzo interviews William Lyne on the topic of UFOs, etc. (https://www.youtube.com/watch?v=k8WOi8YRHvo)
- Nazi UFOs: How they fly (https://concen.org/content/nazi-ufos-how-they-fly-german-tesla-anti-gravity-and-free-energy-program-william-lyne-2004)

The UFOs of Nazi Germany - Viktor Schauberger

- UFO SECRETS OF THE THIRD REICH (https://www.youtube.com/watch?v=IGs9VxqGbkE) – YouTube Video
- Nazi UFOs (https://www.yumpu.com/en/document/view/3882898/viktor-schauberger/91)
- ditto, above (https://1library.net/article/nazi-flying-saucers-ufo-nazi-germany-viktor-schauberger.y9j2pjjq)
- another ditto (https://1library.net/document/y9j2pjjq-the-ufo-s-of-nazi-germany-viktor-schauberger.html)
- PDF file of above (https://www.bibliotecapleyades.net/archivos_pdf/ufos_nazigermany.pdf)
- My mirrored copies (http://vinyasi.info/circuitjs1/texts/Viktor%20Schauberger/)
- download link (https://node1.123dok.com/dt01pdf/123dok_us/004/283/4283957.pdf.pdf?X-Amz-Content-Sha256=UNSIGNED-PAYLOAD&X-Amz-Algorithm=AWS4-HMAC-SHA256&X-Amz-Credential=7PKKQ3DUV8RG19BL%2F20230201%2F%2Fs3%2Faws4_request&X-Amz-Date=20230201T164220Z&X-Amz-SignedHeaders=host&X-Amz-Expires=600&X-Amz-Signature=67c012d5275f9fcf4031ad74b0f7f2a9f71d48e44cfcc19adf4024bc42382c07)
- Source of PDF file, archived (https://web.archive.org/web/20070518074414/http://www.zamandayolculuk.com/Cetinbal/HTMLdosya1/vriltechnology.htm)
- Antarctica's Hidden History: Corporate Foundations of Secret Space Programs Paperback – March 25, 2018 (https://a.co/d/eZaoS7O)

Straying from the topic of UFOs

- Giza's Industrial Complex: Ancient Egypt's Electrical Power & Gas Generating Systems Paperback – May 27, 2022 (https://a.co/d/f88Ef6G)

The Lone Wolf Stigma

In a dramatized biographical movie about Queen Victoria, she says that, "My critics don't have the courage to attack me, directly. So, they attack the people who surround me. I wish they'd stop."

I've seen it happen that if I mention an internet source for my various claims, that source will quietly disappear.

This is the principle of isolation in which my credibility is reduced because I become a lone wolf who is crying to the wind.

Yet, I'll publish these source links, anyway since I don't care if these sources disappear off of YouTube and the Internet.

Reactive Power is not Useless!

Freely available reactive power is never useless, except from a thermodynamic viewpoint, until it is converted (via a resistive heating element) to boil water and rotate a steam turbine to generate electrical energy (as one example of conversion) to do away with nuclear power plants and their byproduct of plutonium.

What's Reversal of Current Good For?

Conventional circuits deplete their voltage source by slowly or quickly equalizing the difference in potential between the two terminals of a fixed voltage source, such as: a battery. They do this by moving a conventional direction of current from higher areas of voltage (occurring at one terminal) towards areas of lower voltage (at the opposing terminal). For example, ...

> A typical 12-volt auto battery will have around 12.6 volts when fully charged. It only needs to drop down to around 10.5 volts to be considered fully discharged.[71]

Unconventional Free Energy circuits, whose current is reversed relative to their polarity of voltage (inducing negative watts as their output power), *increase* the disparity between the terminals of their reactive components, such as: between the two terminals of a coil of wire. Whatever components exhibit this property, these components become the new "sources" of power for these types of circuits replacing (and over-shadowing) whatever contributions may occur from an external source of power.

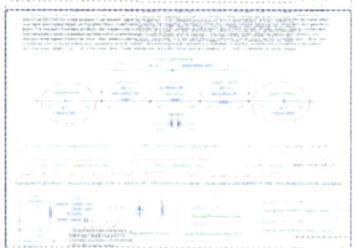

Fig. 2 is a schematic (https://ufile.io/ptgf7eug) of a simulation speculated to be the Ammann brothers' Atmospheric Generator.

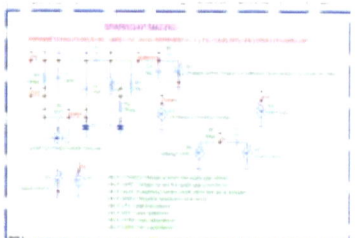

Fig. 3 = Fig. 2 will not simulate verunity without this neon bulb macro from Micro-Cap (http://www.spectrum-soft.com/index.shtm). Its voltage sources are based on conditions of behavior: *if this, then that*.

Fig. 4 = This illustrates the ON/OFF state of the neon bulb, in Fig. 2 (above), and the output of four inductive loads. The escalation of wattage is assisted by an inversion of the polarity of current (relative to voltage) resulting from restricting input and preventing any exit of current.

Explosive Simulation

→ Was the spark transmitter of Heinrich Hertz the Inspiration for the Ammann Brothers Atmospheric Generator? - Quora (https://electricalscience.quora.com/Was-the-Inspiration-for-the-Ammann-Brothers-Atmospheric-Generator).[72]

The top-most graph of Fig. 4 traces the output of a node within the Micro-Cap 12 neon bulb macro (depicted in Fig. 3). This node is labeled "NeonBulb:10" (within the graph of Fig. 4); equivalently labeled "Switchchk" (within Fig. 3), which has already risen from its default value of 10 nano volts to a plateau of 10 volts. This indicates that this neon bulb has turned ON into an arcing plasma.

By the way, if any value closely similar to 10 nano volts were to be traced as the output for this node (within this software macro), then this would indicate a pre-ionizing state preparatory to arcing. This is analogous to what lightning bolts manage to achieve prior to their actual lightning strike. The ionization pathway charts a course preparing for whatever lightning strike may happen to form along this prepared highway.

The second graph (from the top of Fig. 4) traces the output current superimposed over the output voltage of the inductive LOAD as a hyperbolic arch of red dividing the blue underneath. They are diverging at the far right: the red colored current tracing is escalating upwards in the direction of greater positive amperage while the blue colored voltage is escalating downwards in the direction of greater negative voltage. The third graph is tracing the output voltage of the inductive Barrel Coil whose blue-colored arch swerves upwards at an escalating rate of growth in positively signed voltage while the tracing of the fourth graph is red-colored amperage of the Barrel Coil arching downwards at a similar rate of escalation. The fifth and sixth graphs are tracing the rising output of one inductive side of the Copper Tubing while graphs seven and eight are tracing the output of the other side of the Copper Tubing with the neon bulb in between these two halves of copper.

Improving Realism with a Load

I'm still learning.

Up until now, I thought I knew how to "load" an overunity circuit by applying resistance to simulate a mechanical load made upon an inductor (such as: a motor coil). Although this is standard procedure for normal circuits, I was wrong since my simulations tend to produce unconventional triangular waves of inverse polarity between their phases of voltage versus their current. What I see, now — in this *Loaded Test* (to the right), is that a far better method for loading an inductor is to apply a full bridge rectification of four diodes to convert this extreme divergence of phase relations into direct current thereby collapsing the (seemingly) ridiculous levels of output. That's OK. All that is needed (to produce overunity) is to precharge a 100V difference in voltage between the two copper spheres to immediately engage the neon bulb, spark gap, and turn ON overunity in less than ten seconds. Since the simulation errors before we can see what the output levels off at, we won't know whether the output dies out (to zero) after an initial blast, or levels out at some constant output. Blame matrix algebra for taking simulated shortcuts; not "Free Energy".

Loaded Test = When a load (https://ufile.io/8u8yv1sb) is applied to a coil of wire, the infinitely escalating output (of Fig. 4) is delayed.

A Buildable Simulation?

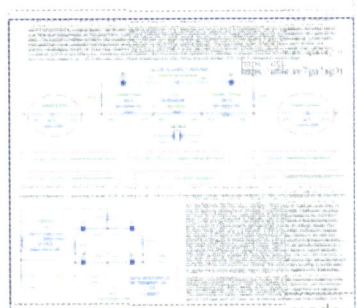

Maybe the Ammann brothers made use of baking soda diodes (https://ufile.io/7ga7ag3t) to streamline the performance of their capacitors and make them safe?

High voltage tolerant diodes of a century ago were similar to electrolytic capacitors of today. They were made with an aluminum cathode which was conditioned by subjecting it to alternating current. This would form a layer of aluminum oxide on top of the aluminum. When this "conditioned" plate of aluminum was immersed in a watery solution of either borax or baking soda, and paired with an anode of some other substance, other than aluminum, then a diode function was fulfilled with a large tolerance to elevated voltages.[73] [74] Salt would have made a better electrolyte if this were truly a capacitor and not a diode, but with the risk of producing chlorine gas along with hydrogen and oxygen gases all of which are *very* explosive if not vented away from us into a safer area.

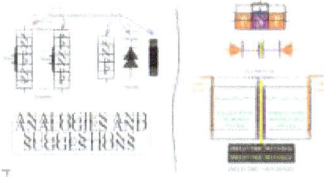

These are analogies and suggestions of transistors versus diodes in the schematic to the left.

Ammann brothers' original newspaper photograph.

Pairing these diodes with a capacitor sandwiched in between them and with both of their cathodes pointing towards each other, and towards their shared capacitor, helps to reduce the need for simulating a very small capacitance (of 1e-21 Farads) possessing a very large equivalent series resistance (of 1e+8 Ohms)[75] inside of the simulated capacitance, labeled: *ProximityOfSpheres* by comparison to if this capacitance had been there all alone without any diodes to assist it. This design may be more likely to be buildable and, thus, more practical to take seriously.

By the way, ...

I suspect that this specially designed capacitive diode, with both of its capacitor plates consisting of oxidized aluminum immersed in a non-electrolytic bath of borax or baking soda, sat behind -and connected to- the headlight sockets of the Ammann brothers' batteryless EV serving as an additional capacitance with respect to the air-gap between their two copper spheres. So, I am guessing that their two hollow copper spheres were serving as capacitors of their own (and doubling as aerials grounded to the electrostatic charge of the atmosphere) and shorted out with this high-voltage tolerant capacitor hidden inside the front-end of their car which may have served as a throttle?

There still remains the issue of a lack of regulation to prevent the simulated behavior of an explosive force. Whenever the spark gap (neon bulb) flashes ON, the output of power takes a sharp ascension. Whether or not it's merely a momentary surge, I wouldn't want to be around when it explodes!

Does someone want to risk their life by testing out my theories? Because these simulations are suggesting the electrical equivalence of a bomb!

An Example of Domestic Terrorism

is.gd/disruptgrid

Disrupting the Power Grid

Fig. 5a, schematic (https://ufile.io/7sidu1df) – Watch out! Don't position your free energy device within city limits unless it is fully shielded to prevent it from disturbing radio transmission nearby.

I don't think it is entirely fair that C. Earl Ammann was charged with "stealing energy from the electric utility grid" when he arrived in Washington, D.C. to deliver his EV to the United States Patent Office when the grid already steals from its surrounding environment. He and his brother had demonstrated their EV on the streets of Denver, Colorado, driving it around town, up and down hills, while running it without any batteries.[47] [3]

Fig. 5b – Energy disruption on the grid.

This arrest is why you and I never heard of him until I ran across a few people on EnergeticForum (http://www.energeticforum.com/forum/energetic-forum-discussion/renewable-energy/14490-ltspice-simulation-of-electronic-boost-via-the-isolation-of-voltage-current-sources#post435066) talking about him. And, now, you know a little of his story.

His disruption of the electric utility grid does not mean that he was a fraud. Oh, contrair! It means that the grid got in the way of the over-reactance of a free energy circuit (aka, the negative impedance of Foster's reactance theorem) for having its reactance increased in a favorable way and at an alarming rate without the addition of any more energy. The consequence, of this over-reactance, was that an abundance of magnetomotive force was inductively transmitted to the grid. My simulations (in Fig. 5a and Fig. 5b) indicate that a lot of current was being *pumped into* transmission lines located nearby and *into* the wiring of the homes of his immediate neighbors. Yet, his device would still have been overunity, without a grid nearby to pump energy into, had they located themselves out in the middle of the ocean or in the desert or on some lonely mountain top. His device would not have escalated its power level so fast without the grid to have a convenient place to drain its excessively negative impedance into by an over-stimulation of the rate of reactance per unit of time. But, regardless of the rate of escalation of power within his device, it would still be overunity.

As my simulation indicates, up-above, in Fig. 2 (which presumes living out in the countryside far away from the electric utility grid), this style of free energy circuitry performs very nicely without any help from energy sources, nearby, getting in their way. Their over-reactant, negative impedance comes from within themselves upon their scant stimulation from external sources of energy, such as: the ambient charge existing in the atmosphere at ground level. This amounts to a mere micro volt which is amply sufficient for stimulating over-reactance in a circuit of appropriate design.

But regardless of whether a free energy circuit gets its incentive for the movement of its energy from a power source, or else gets it from the natural environment, or else gets this incentive from within itself, either way, we have no way of performing a controlled experiment to make a clear determination as to where does this extra incentive come from since we won't be able to isolate any possible answer so as to avoid confusion or error. Hence, we'll just have to assume nothing until we find out. Otherwise, we'll be making ourselves look very overbearing and pompous in hindsight!

Since this style of circuitry does not require an external power source, but does require an external catalyst of stimulation, care must be taken to restrict external sources of power from the disruptive influence of this type of circuitry to protect them from becoming burdened with excessive disruption coming from this highly reactive type of circuit. Over-reactance can become a sponge of inverted current that sucks additional energy from out of nearby sources of energy if allowed to do so without human intervention. Or, over-reactance can become a pump which performs the opposite task of sending extra energy *into* nearby sources (or conduits) of energy, such as: transmission lines, against their impedance and/or resistance.

Oh, pooh!

It is this incentive, born of reactance, which supersedes prime movers into becoming mere catalysts while the prime movement shifts to the over-reactance which is spawned by this style of circuitry. It is the inversion of current, also known as: negative impedance, which broadcasts its influence into its environment including manmade sources of power, such as: the utility grid. This is why over-reactance has been the bane of electrical engineers, for there are two sides of reactance, either: benevolent or demanding. We have to take care to restrict our use of reactance to benefit our appliances without destroying our sources of energy in the course of utilizing them. We do this by becoming mindful of the fact that we no longer need a source of power to fund our devices. All we need is for those sources to catalyze an over-reactance. Once over-reactance takes over (if we let it), it -then- becomes the dominant source for the accumulation of proto-energy (radiant energy; current inversion) which can -then- be converted into real power through mere (positive) impedance and resistance, alone.

Our sociological "motivation for profit" must be restricted to our motive for leading a productive life without allowing this "motive for profit" to unduly burden anyone or anything. So, I am advocating efficiency and the fair treatment of the consumer in the course of pursuing "free energy". Profit has become the bane of the consumer especially in the wake of inflation in which *profit becomes inflated* making its pursuit an automatic infringement upon human decency.

> The very foundation of our society has been predicated upon the profit motive. Yet, its pursuit has spawned the inflation of our economy making its continued pursuit a violation of human dignity and welfare.
>
> There's no profit to be made from "free energy" if no one can charge us for its consumption.
>
> We can avoid being charged for our energy usage by recycling its electrical reactance to such a degree of excessive **conservation** that a mere factor of 99% reuse (for instance) constitutes a 100 to 1 gain (of output versus input) without any violation of physics.[76]

Non-Explosive Simulation

is.gd/stableOU   **OR**   is.gd/stableou

Stable, non-explosive schematic (https://ufile.io/ks2boanx) fed from a 1½V solar panel (or battery) giving a plateau of 900k Watts, RMS, building up its output over 300ks of run-time (83⅓ hours ≅ 3½ days).

It's possible to achieve a non-explosive simulation (https://ufile.io/ks2boanx) of a stable output in Micro-Cap 12 electronic simulator (http://www.spectrum-soft.com/index.shtm). Power is enhanced if the two ferromagnetic coils, Armature1 and Armature2, possess slightly different number of windings of insulated floral (green) iron wire in contrast to each other. This simulation was hosted on a 64-bit computer. Hosting Micro-Cap 12 on anything less than this, such as: on a 32-bit computer, tends to give (what may be considered) a greater tendency for false positives with regard to overunity of gainful output.

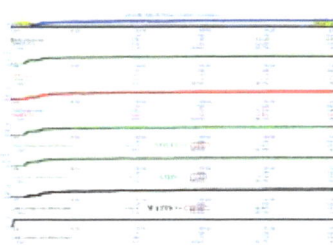

Stable, non-explosive output.

Conventional vs Non-Conventional Circuits

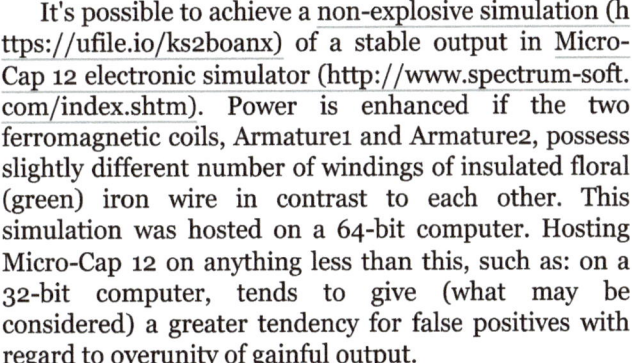

Fig. 6a compares two circuits: the output on the bottom graph displays an explosion of gain due to the inversion of current arising from the plasma state of a neon bulb, spark gap.

→ Lest anyone entertain the erroneous opinion that all of this is due to imperfections of simulation due to round-off error,[77] [78] here are a pair of variations of a conventional circuit (Fig. 6a and Fig. 6b) which costs a conventionally high drainage made upon some batteries. Yet, the magic does not get initiated there. True, they will contribute their excessive drainage of current, but the magical orchestration will originate (not at the batteries, but) at the neon bulb turning ON (arcing into a plasma) when it reaches its breakdown of resistance at, or above, 90 volts (which is what Micro-Cap simulator sets this threshold at). Then, and only then, will the batteries exceed their prior drainage of nearly 450k amperes to achieve an escalation quickly rising to infinity! This demonstrates the magic of the inversion of current (relative to the phase of voltage) arising, here, exclusively from the neon bulb rather than from any fancy arrangement of electrical components (usually: inductances and capacitances). All of the costs of energy to run this simulated circuit are conforming to conventionally high values expected of them so as to minimize the possibility of no one taking this example, seriously.

https://ufile.io/856ymhga

Fig. 6b is a more efficient method (https://ufile.io/856ymhga) of flashing a neon bulb ON instead of a conventional method which needlessly wastes input.

The example on the right is more efficient along a style that I frequently employ of using precharged capacitors and/or voltage sources rated at around 1μ volt (a sine wave generator in this example). In this case, this sine source provides a very important frequency of sufficient pitch to accelerate an opportunity for an explosive gain of amplitude to occur without wasting a whole lot of power to facilitate this opportunity. The power is provided by the 10 Farad capacitors (possessing a maximum of 400 milli Ohms, each, of equivalent series resistance) and each are precharged with 100 volts of opposing polarity to coincide with each other in their circular arrangement.

Sometimes, it's important to distinguish between frequency and power and separate them so as to not waste a continuous stream of power to maintain a frequency. This frequency can be very useful in accelerating the time it takes for reactance to explode and yield significant results of amplitude despite the fact that neither formula for electrical reactance (inductive or capacitive) has any factor of kinetic energy, such as: power, amps, or volts, inside of it. Instead, they possess potentialities of power, such as: frequency, inductance, and capacitance per cycle of oscillation defined in terms of angular momentum, or: 2π. Here is another reality to energy which is often overlooked regarding the inherent potential energy already resident within a circuit, namely: its momentum.

Thus, if we focus on a circuit's momentum, rather than focusing on giving the circuit any more energy in addition to whatever it already possesses, then we have an opportunity to manipulate this momentum using the potentialities of: frequency, inductance and capacitance. This does not cost us any more energy than what has already been fed into our circuit.

Think about it ...
Isn't this focus on momentum the foundation for anti-gravity levitation? And doesn't electrical reactance make inertia equivalent to gravity?[79] How else do UFO's stay aloft? And make right-angle turns at high speed without slowing down? And suddenly stop without deceleration?

This makes me wonder if we have overlooked a very significant perspective in both physics and electrodynamics. Hmmm, ...

No one (usually) thinks of the mass of a coil as possessing potential energy unless that someone was Joseph Newman. Doesn't matter what people thought of him or his ideas of gyroscopic power (http://www.rexresearch.com/newmanpatents/newman2.html). Maybe that was his way of describing inductive reactance? What matters is that, at least, he understood the potential power which is inherent within the inductance of a coil and made use of that power even if it could've been done in a more efficient manner. It almost doesn't matter. At least he confronted people with working models even if he may have lied (in his book (https://isbnsearch.org/isbn/9780961383527)) on how to build it.[80]

Remember, ...
Input power must step aside and quickly dissipate (using standard thermodynamics) to reduce input and, yet still be able to maintain an excellent output. {I could have used the word: 'conserve' instead of 'reduce', but that might confuse anyone who is brainwashed to think of the laws of physics instead of the economics of conserving our electrical resources.}

Negation of current is a powerful factor, within overunity circuits, since negation of watts and the divergence of voltage differences (between two nodes within a circuit) are the result. This leads to the non-saturation of current within inductors (exhibited by triangular waves, or spikes) and a continuous escalation of power at an exponential rate. This rate may not be constant! In other words, a nicely smooth hyperbolic (ascent or descent) away from an oscilloscope's midline of zero may suddenly become a vertical slam into infinite gain!

Similarities to Eric Dollard's LMB Analog Computer

Fig. 7a compares the pros and cons of two very distinct forms of electrical transmission: the normal type which suffers intense losses versus its converse which gains momentum!

Fig. 6b, up-above, and Fig. 7b (on the right) possess stark similarities to Eric Dollard's Analog Computer (https://www.youtube.com/watch?v=6BnCUBKgnnc) in Longitudinal Magneto-Dielectric (https://www.youtube.com/watch?v=n8drff4j90) (LMD) mode since those circuits exhibit their dielectric force (measured in voltage) across their vector of transmission (in series to) their magnetic force of support (in parallel with) their vector of transmission (Fig. 7a). Hence, Eric has managed to create a whole new orientation of transmission existing in the space between a pair of transmission wires in which each whole wire is one of two terminals of transmission while the space between these two terminals is the line of transmission. Since this line of transmission is empty space, this constitutes a line of longitudinal dielectricity while each terminal is a solid composition of magnetizable transverse conductance.[81]

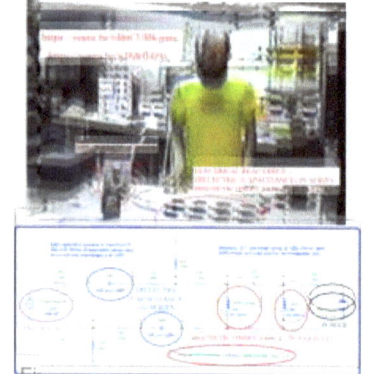

Fig. 7b – LMB analog computers.

This is, actually, more efficient at transmitting energy since the magnetism of each terminal remains where it is initially located and we do not attempt to move it anywhere. This is a great boon since we have learned, from studying history, that the movement of magnetizable current (in the trans-Atlantic telegraph cable problem of the 1800s) drops off very rapidly over distance making that conventional style of transmission very costly. Instead, we polarize the empty space between a pair of transmission wires with a capacitant charge of voltage (using each magnetizable terminal of conductance as a sort of capacitor plate to this new style of transmission).

Capacitive Negative Resistance Suggests Epicyclic Lunar Rotation

At 9 minutes and 20 seconds into this YouTube video (https://www.youtube.com/watch?v=ghVubdMIMVM&t=560s), we are shown the inside of an electric generator which reminds me of the inside (http://gp.alternate-energy.net/john-bedini-30-coiler-energizer-e86025etd.html) of John Bedini's patented energizer (https://patentimages.storage.googleapis.com/7a/96/8a/dd335ecc71e09/US20030127928A1.pdf) which is different than standard rotary generators.

Fig. 8a – Negative resistance will grow in amplitude (over time) if it is dielectrical, namely: located within capacitors. Negative resistance will shrink in amplitude (over time) if it is inductive or conductive, namely: located within inductors or resistors.

Fig. 8b – Negative resistance (https://ufile.io/892a7ysb) can vary its behavior based on its location.

Here are (http://gp.alternate-energy.net/john-bedini-30-coiler-energizer-e86025etd.html) two more examples (http://johnbedini.net/tc.jpg) of Bedini energizers.

This puts into doubt that the moon is orbiting the Earth since its axis of rotation is not centered within its own (lunar) center of mass but is centered within the Earth's center of mass making it appear as if it is not in orbit around the Earth. Instead, the moon's rotation around the Earth is an extension of the rotation of the Earth's center of mass.

The reason why I call this simulated experiment *moon* (in Fig. 8a and Fig. 8b) is due to John Bedini's design for a very efficient rotary generator and how the moon always keeps the same side facing us throughout its orbit around the Earth. If the moon (representing a coil) had rotated while orbiting the Earth, then it would have been engaging Lenz's Law (https://www.electrical4u.com/lenz-law-of-electromagnetic-induction/#:~:text=Lenz%E2%80%99s%20law%20of%20electromagnetic%20induction%20states%20that%20the%20induced%20it.) as a consequence of Michael Faraday's Law in which the movement of one magnetic reference frame against another magnetic reference frame produces a counter-opposing force known as: back EMF (https://pressbooks.online.ucf.edu/phy2054/chapter/back-emf/#:~:text=Back%20emf%20is%20the%20generator%20output%20of%20a,current%20when%20it%20is%20on%20but%20is%20not%20turning.). This undermines motor and generator efficiencies. Since movement is occurring between the magnetic reference frames of both coils (in these simulated examples, of: Fig. 8a and Fig. 8b), this suggests negative resistance is being engaged within the context of inductive and conductive reactances diminishing the amplitude of the waves which are traced in the middle pair and bottom pair of the oscilloscope tracings of Fig. 8a.

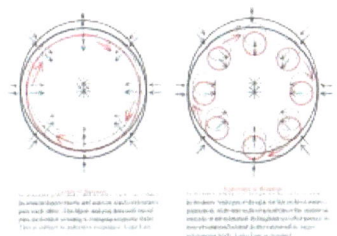

Fig. 8c – compares cyclic versus epicyclic rotations of standard generator/motor designs versus John Bedini's suggesting lunar rotation around the Earth (in Bedini's case):

But if the moon's spheroidal inductance rotates around, not the center of its own mass, but rather the center of the Earth's mass; then it maps an epicycle path of rotation which does not engage Lenz Law since the surfaces of the inner "red" coils do not move relative to the surface of the outer "black" coil in Fig. 8c. They move together much like a homopolar generator. Since the smaller "red" coils on the inside of Fig. 8c are not rotating relative to the larger and external "black" coil encircling the inner coils, that is one less type of movement to amplify back EMF. Hence, efficiency is improved. This suggests (to me) that negative resistance is engaging dielectric (capacitive) reactance graphed in the top pair of tracings in Fig. 8a.

What does this mean?

Effectively, it means that the moon has no mass and, thus, no angular momentum of its own apart from that imparted to it by the Earth.[82] For if it did possess its own angular momentum, then it would rotate around its own center of mass. Yet, it does not. Instead, it rotates around the Earth's center of mass and will continue to rotate only once around the Sun's center of mass for every cycle of its orbit (once its perturbations of sudden release died down) if the Earth were to suddenly disappear. Mass has the consequence of generating inertia. Without a mass to generate inertia, the moon is "locked" into the Earth's center of mass. Hence, *did the Apollo lunar landings actually occur? Could they have occurred? Yes. How?*

Because this relocation of the moon's center of mass is performed by the moon being composed almost entirely of aluminum, or something similar, which possesses the property of paramagnetism, and if the moon is also largely a vacuous object of a lightweight, but structurally sturdy and honeycombed construction. These stipulations deflect its own center of mass to the Earth which is largely ferromagnetic (due to the Earth's preponderance of iron in its makeup) which has the property of consolidating its own center of mass plus requisitioning any other object's center of mass to add to its own.

If the moon has any mass (and it's safe to assume that it does, or else the lunar landings would never have occurred ;-), then its mass has been neutralized. Thus, a buoyancy exists which suspends the moon at its height above the Earth instead of our presumption of the mutual gravity between the Earth and the moon acting in conjunction with our assumption of its possession of an orbital momentum as it traces a pathway around the Earth. But under the circumstances, the moon neither possesses orbital momentum nor does it possess angular momentum to rotate. Both actions are taken care of by the Earth's angular momentum and its largely iron composition.

Capacitive negative resistance suggests the epicyclic rotation of our moon since resistance would have to be gainfully negative in order to compensate for the moon's lack of mass with respect to its size. Any other type of resistance would require (demand) a significant mass within the moon to give it sufficient momentum to rotate independent of its orbit around the Earth. Since its orbit is epicyclic, then it cannot have any significant mass. It is a vacuous sphere.

Conversely, this is why the Earth and our Sun are both composed, mostly, of iron. According to the website: TheSurfaceOfTheSun.com (http://thesurfaceofthesun.com/) – the Sun's hard surface is composed of calcium ferrite underneath its atmospheric ocean of neon and silicon plasma.

How does Free Energy not Violate Conservation?

The difference between reactance and energy is that reactance can be disassembled into its constituent ingredients and avoid Conservation of Energy (since reactance is not energy) while energy cannot. Energy is a singularity while reactance is composed of a multitude of factors. When we analyze the factors of electrodynamic energy, such as: voltage and current, we are analyzing our experiential expertise of the factors of reactance and mapping their equivalencies to the electrodynamics of energy which we cannot disassemble (to analyze) no matter how much we would like to. But, we can disassemble reactance since it contains analogous components of capacitance and inductance which directly map to their storage of voltage and current and possess their physical counterparts of capacitors and inductors, respectively.[83]

It is important to make these distinctions so as to remind ourselves of this difference between reactance and electrodynamic energy and how useful these analogues are to our analysis of electrodynamics to avoid getting lost and confused by their translations. If we allowed ourselves to disassemble electrodynamic energy in the same manner in which we disassemble reactance, we would violate Conservation of Energy.

See the point?

A figure skater in a spin uses conservation of angular momentum – decreasing her moment of inertia by drawing in her arms and legs increases her rotational speed.

Definition of Moment of Inertia

For a simple pendulum (an oscillating body/system/circuit), this definition yields a formula for the moment of inertia I in terms of the mass m of the pendulum and its distance r from the pivot point as, $I = mr^2$.

Thus, the moment of inertia of the oscillating body (pendulum) depends on both the mass m of a body and its geometry, or shape, as defined by the distance r to the axis of rotation.

How do we accomplish an exponential rise of potential energy without violating the conservation of angular momentum? By separating the mass m of the moment of inertia from its radius (squared) r^2, we dematerialize the moment of inertia into its reactive equivalencies of: inductance H representing mass m, and capacitance F equaling radius (squared) r^2. But we may only accomplish this whenever current becomes inverted relative to voltage. And this *dematerialization* occurs due to a translation from real numbers dominating the scene into complex numbers taking over with emphasis on its imaginary coefficient $\sqrt{-1}$ dominating the complex number field of the reactance of an electrical component.

When we succeed at reversing current, then there will be no delay: there will be no storage delay within a coil of wire and there will be no storage delay within a capacitor. And there will be no delayed response whenever *virtual* momentum is stored versus whenever it is released. Coils and capacitors, at this point, become mirrors which merely reflect without storage. Since current is inverted, then there is no delay, and coils and capacitors are receiving the current aspect of their power at the same time that they are exporting the voltage aspect of their power. Hence, a capacitor no longer behaves strictly as a capacitor and a coil of wire no longer strictly behaves as a coil of wire. Each begin to take on the characteristics of the other, but merely in a dynamic manner.

This dynamic condition creates a transformation of the usual dictum of physics, in which: "Energy IN has to equal Energy OUT" becomes true, not for the entire circuit, but for each and every component due to the reversal of current. This reversal of current eliminates any time-delay that Newton's Law of Reaction for every Action would assume.

Furthermore, this reversal of current assumes that what was true in the prior half-cycle of oscillation is no longer true in any subsequent half-cycle due to this separation between voltage and current of one-half cycle of angular displacement (in time; per cycle).

In other words, ...
The result of the previous half-cycle becomes the input for its subsequent half-cycle. And since current is reversed, mass has become separated from radius (squared) r^2. This fact, alone, severs any relationship between this process of Free Energy magnification and the conservation of angular momentum across multiple half-cycles.

In other words, the conservation of angular momentum is, now, only true for each half-cycle of oscillation while no longer being true across two or more subsequent half-cycles. This is due to the constantly changing features of moment of inertia occurring between any two subsequent half-cycles of oscillation.

Noether's Theorem allows for this discrepancy when it states that the loophole for the Conservation of Energy is whenever time-frames undergo alteration, because conservation is assumed to be true exclusively within the same reference frame (for time); not across two separate and distinctly different time-frames of reference.

> *The energy conservation law is a consequence of the shift symmetry of time; energy conservation is implied by the empirical fact that the laws of physics do not change with time itself. Philosophically this can be stated as "nothing depends on time per se". In other words, if the physical system is invariant under the continuous symmetry of time translation, then its energy (which is the canonical conjugate quantity to time) is conserved. Conversely, systems that are not invariant under shifts in time (e.g. systems with time-dependent potential energy) do not exhibit conservation of energy – unless we consider them to exchange energy with another, an external system so that the theory of the enlarged system becomes time-invariant again.*

> Editor's note: *"nothing depends upon time per se"* – Someone went *to sleep at the wheel* while driving their proverbial electrified vessel! *Apparently*, physicists could care less about electrodynamics in which electrical reactance depends *heavily* upon time as its foundation since reactance (ie, *"time-dependent potential energy"*) has no dynamic outside of time. Hence, time-frames (ie, cycles and half-cycles of oscillation) *matter a lot!*

> By the way, ...
> This is not quantum mechanics in which "black holes" and "time travel" needs to be invoked for the reversal of light (acting as current reversal) to occur. Instead, simple electrodynamic theory applies, here.

It so happens that the reversal of current satisfies this loophole in as much as no two *half-cycles of oscillation* share the same (equivalent) time-frame. Only each *half-cycle* of oscillation can be said to be true to its own time-frame servicing its own reference-frame.

And, ...
Electrical reactance formula (used for calculating the inductive and capacitive reactances of inductors and capacitors) bridges the time-frame gap existing between (and across) multiple half-cycles of oscillation since each iteration of calculations of reactance are always true per half-cycle of oscillation, but not true for the next half-cycle since a distinction must be made (in time) between the inductive reactance resulting from one iteration of calculation (from the formula for inductive reactance using the inductance of the magnetic field of an inductor) from the inductance of the prior half-cycle which spawns the inductive reactance (namely: the inductance) for the subsequent half-cycle. Likewise, this is true for calculating capacitive reactance versus the capacitance which spawned it.

Ergo, due to the reversal of current, there can no longer be any distinction made between inductance and inductive reactance. Nor can there be any distinction made between capacitance and capacitive reactance, for over time: these distinctions which we used to hold so dear in a static world of make-believe conditions of stability of time-frames is no longer valid outside of any singular half-cycle.

Not until current reforms back into its normal relationship with voltage (in which the phases of oscillatory current are in alignment with the phases of oscillatory voltage) will a whole new value of angular momentum materialize *literally out of thin air* (out of the reactances of counter-space). Only, then, will conservatives cry, "foul play".

But if we bypass the jurisdiction of the Conservation of Angular Momentum, then no law has been violated!

So, why all the fuss?

By dismantling time, we dismantle conservation. This is what reversal of current manages to accomplish. But only if it is accomplished via analog components; not digital components.

True, ...

I've had to use a digital medium of computer simulations to come up with these conclusions and insights. But that's because I trust these multi-thousand dollar simulation softwares are honest in their appraisal of electrodynamics.

And they are honest.

Besides giving me an unadulterated view of electrodynamic theory, they also (sometimes) honestly let me "in" on their petty little secrets regarding their policy to tweak whatever their software designer thought was wrong with electrical reality by sometimes "fudging" the software's results. Such as: limiting the current of a diode should it rise above 1kA. This began to bug me until I could no longer tolerate this behavior. This motivated me to peer into the software code (of the simulator in question) to discover a comment made by its designer that: "sometimes, diodes act weird".

To me, that is not acceptable to get a degree in electrical engineering from a prestigious university only to fudge a diode's behavior, because of finding personal fault with it!

In Conclusion: *What is electricity?*

is.gd/refineohmslaw

If I rephrase the question as …

What is electrical power, then the correct answer is to say that Ohm's Law is a combination of two components.

The first component of electricity is real voltage which is distributed across space. We will label this type of voltage with the label of: V_r, to signify that this represents Real Voltage.

The second component of electrical power is reactive voltage existing in time. This latter component is divided by one or more various impedances, tempting us to simplify this second component of electrical power by way of mathematical substitution in which a singular symbol, I, called: "current," replaces reactive voltage divided by impedance. This latter, more accurate version of the "current" portion of Ohm's Law can be signified by: $V_x \div Z$.

Hence, Ohm's Law fails to describe power (P, watts) as…

$$P \equiv \frac{V_r \times V_x \times \sqrt{-1}}{Z} = \frac{V_r \times jV_x}{Z}$$

…if we also assume the substitution of j representing the square root of negative one whenever utilized within the field of electrical engineering: $j = \sqrt{-1}$, so as to avoid confusion with the letter I used to represent *current*.

Instead, conventional wisdom allows for their equivalence…

$$P \equiv \frac{V^2}{Z} \equiv V \times I$$

…but fails to distinguish among types of voltages and the implications of expanding our consideration of reactive resistance, namely: impedance Z. This mathematical shorthand suggests the illusion that voltage is squared and then it is divided by resistance due to the illusory temptation to assume that there is only one type of voltage rather than two.

Yet, we know that there is electrical reactance within all types of circuits to one degree or another. This awareness is predicated upon the fact that a piece of wire (for example) exhibits inductive reactance along its length and capacitive reactance extending radially outward from its center across its surface (if it's merely bare) plus across its insulation (if it has any on its surface). Thus, a simple flashlight circuit possesses electrical reactance. Yet, this reactance is so minor that we tell ourselves that we may safely ignore it without worrying too much about making some sort of blatant error.

But this will only work some of the time. We cannot guarantee that this will work most of the time, much less all of the time. And it will certainly never work out very well within the context of my style of orchestrating electrodynamic behavior.

It is this sort of mental programming that all of us must confront (at one time or another) when we wish to expand our awareness of electricity in general and free energy in particular.

We also know that voltage drop cannot be performed upon imaginary numbers.

This temptation to simplify Ohm's Law makes the job of the technician vastly easier to follow procedures laid down by policies which encourage the monopolistic belief that "there is no such thing as a free lunch."[84]

But if we assume a scarcity of freely available input power, then we are in a much better position to favor over-reactance as a superior source of renewable energy.

Appendix

Photons do not Exist

It is not necessary to theorize the anomalous existence of photons to account for the traversal of energy across empty space when space is perfectly capable of acting as a dielectric medium supporting the existence of longitudinal shock waves.[85] Electromagnetic transverse ripple waves are a short-range ramification of dielectric (ie, electrostatic) longitudinal shock waves converting into transverse ripple waves whenever longitudinal shock waves meet up with matter at the other end of an empty void of space.

Oliver Heaviside effectively acknowledged this, over a century ago, when he devised his Telegrapher's equations to solve the riddle of: "Why was the magnetic field of electricity dying out so rapidly (along the length of the trans-Atlantic telegraph cable) while the electric field did not?" It was because the electric field does not travel since it is the consequence of a dielectric material responding to the imposition of a potential storage of voltage. This dielectric material was the boundary condition initiated by the surface of the copper cable separating the cable from its surrounding space of Atlantic ocean. The insulation of this cable helped facilitate this boundary condition insuring that no current would leak out into the ocean. But the dielectric condition of a transmission cable is at right angles to its transmission while its transmission is parallel to the cable's length.

1858 trans-Atlantic cable route.

So; :::

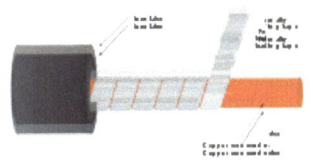

Permalloy loaded cable construction. Compare with: Pupin's coils (https://sr-m-wikipedia-org.translate.goog/wiki/%D0%9F%D1%83%D0%BF%D0%B8%D0%BD%D0%BE%D0%B2_%D0%BA%D0%B0%D0%BB%D0%B5%D0%BC?_x_tr_sl=auto&_x_tr_tl=en&_x_tr_hl=en&_x_tr_pto=wapp) for paired telephone transmission lines (image to the right).

The transmission of dielectric charge of potential (voltage) does not have to travel, unlike the magnetization of current which does travel. Thus, voltage potential does not have to die-out while magnetic current *must die-out* along the entire length of a copper cable due to the resistance which copper conduit offers to the flow of current. Hence, a ferromagnetic wire or tape had to be wrapped around the bare copper cable before applying a very thick layer of insulation to retain the magnetic field (generated by the application of a difference in voltage potentials upon the terminals at either end of this copper cable) to prevent the loss of the magnetic field surrounding this cable.

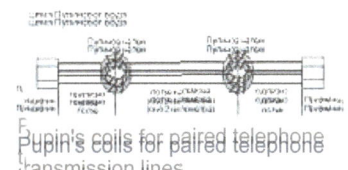
Pupin's coils for paired telephone transmission lines.

This prevention of the loss of magnetism is known as: magnetic remanence, or simply: remanence. It was used as a form of computer memory between the years of 1955 and 1975 by creating tiny ferrite rings through which was threaded a pair of crisscrossing copper wires creating a tapestry of horizontally and vertically aligned fine copper wires each of whose intersections was surrounded by a single, magnetizable, ferrite ring which could remember the polarity of its magnetization long after the application of voltage was shut off in the wires which had passed through each ring. This polarity of remembrance was interpreted as either a binary "one" or a "zero". And this memory of ferromagnetic material is perpetual. It never dies out unless acted upon by a new force of electricity. This is in keeping with Sir Isaac Newton's dictum, that: "Energy tends to remain in a particular vector of motion unless acted upon by another vector."[86]

You see, :::

Magnetic energy is a preexisting condition within a ferrite ring. All we do is make use of it by organizing its random polarizations into a collective alignment which we can recognize as possessing a north and a south pole held to be en masse across the entire chunk of this ferromagnetic material.

So, the perpetuity of magnetism is already within the ferrite ring. But it's a chaotic mess until we impart order to it and, thus, put it to work for our benefit.

It is this perpetuity of ferromagnetism within a lengthy strand of permalloy (or similar) tape which made the transmission of current possible across the trans-Atlantic telegraph cable – without which, there would have been no Morse coded message received.

It is only short lengths of copper wire which can carry a magnetic charge. Long lengths don't succeed unless ferromagnetic material is located nearby, or else this lengthy copper wire is coiled so that the leakage of one turn of wire leaks out into the next!

So, ...
What is the boundary condition of space which makes the longitudinal transmission of dielectricity instantaneously possible?

The answer is, ...
The existence of matter at either end of a longitudinal transmission is what makes this transmission possible across empty space. This space acts similar to the behavior of a dielectric material sandwiched between two conductive plates within a capacitor. And the boundary condition of two conductive plates (on either side of a capacitor's dielectric middle-layer) respond to the longitudinal transmission across the dielectric material by creating transverse ripples of current at the conductive plates located on either side of this dielectric sandwich. But these ripples of current are short-range dying out very quickly due to the resistance of the conductive material in which they arise unless this material incorporates the use of a ferromagnetic mass, such as: iron, or a coiled geometry of the copper coil, or both, to help "remember" the magnetic ripple.[87]

A 32 x 32 core memory plane storing 1024 bits (or 128 bytes) of data. The small black rings at the intersections of the grid wires, organized in four squares, are the ferrite cores.

Electrons do not Exist

- Electricity is not "electrons", if it was, how could particles related to current (closed circuit) flow in a single wire? (https://electricalscience.quora.com/Electricity-is-not-electrons-if-it-was-how-could-particles-related-to-current-closed-circuit-flow-in-a-single-wire?ch=10&oid=86886185&share=3bf12aab&srid=3zXXZ&target_type=post) – This is a *good question* posed by Franco Bruno Corteletti at Quora.

Conservation of Energy does not Exist!

Unless you like to believe in fictions! ⇒   is.gd/conservfict   **OR**   is.gd/conservationisafiction   **OR**   is.gd/conservationofenergyisafiction

Three fictions, to date, have I discovered corrupting electrical engineering with their shams! They are posted on Quora (https://electricalscience.quora.com/Three-fictions-to-date-have-I-discovered-corrupting-electrical-engineering-with-their-shams-They-are-1-Current-2?comment_id=44543174&comment_type=3) ...

1. Current.
    1. Current is a mathematical fiction constructed out of convenience to replace its more complicated counterpart of "reactive voltage divided by impedance".
2. Conservation of Energy (the audio posting, below, on Podbean).
    1. Likewise, Conservation of Energy is a geometric and mathematical construct born of the mind of man, derived from Kirchhoff's Current Laws pertaining, strictly, to the nodes in between circuit components. Although convenient for use in calculating the expected behavior of a circuit, nodes do not exist in physical reality, and you won't find any physical proof for their existence. Since Conservation of Energy is the direct consequence of Kirchhoff's Law for Current computation, and these computations are strictly performed upon fictional nodes (not upon the components which lie in between fictional nodes), both nodes and the Conservation of Energy are guilty of promoting themselves in the name of, and masquerading as a substitute for, their physical counterparts. So, I see no intrusion by physics into this misrepresentation unless it is a fault of physicists for originating this misconception. I don't know. Someone is responsible and it doesn't matter who at this point since all of us are guilty of maintaining it.
    2. Conservation of Energy is also a fiction due to the non-isolation of any system of energy you could possibly think of! It was never intended as anything other than an hypothesis to assist the physicist, or the engineer, in assessing the energy accountability of energetic systems. It was never intended to also be applied to the real world. Yet, don't blame the United States Patent Office for imposing a fiction upon all wannabee inventors. They're just following orders! Whose orders are they following?
3. Reactive Power.
    1. Putting the words, Power, and, Reactive, together in the same sentence is like mixing oil and water. They are about as unalike as can be. Power is something real that you can sense with any of your five senses while reactance has to be taken on faith and the testimonials of countless engineers who have come before us for the past century and a half.

2. But testimonials do not, cannot, substitute for lack of physical proof.
3. Power is something real, while reactance is something imaginary.
    1. Real, as in: real numbers; Imaginary, as in: the square roots of negative numbers.
    2. Yet, reactance is the *cause* of the expansion or contraction of power. It is not restricted to merely the reactive *effect* of the application of power which is made upon a reactive component, such as: capacitors and inductors.

So, reactance is the cause of the expansion (not the creation) and contraction (not the destruction) of power, but only if we divorce the concept of power from any consideration of reactance. It's a four-step process of effectively promoting this sham. They are ...

1. Create the sham of *current* from out of the reality of "reactive voltage divided by impedance."
2. From the sham of *current,* create another sham called *Kirchhoff's Current Law* and call it a law when Eli Pasternak, an electrical engineer on Quora, admits otherwise:[11] "Kirchhoff's laws are only an approximation for the purpose of simplifying circuit design. They are not exact laws. For instance, they ignore the magnetic fields in the conductors of a circuit and the possibility of mutual inductance of adjacent circuit loops. In this simplified model, energy conservation is merely a statement that all the electric energy is passing through a node and thus it must be conserved."
    1. BobD (https://physics.stackexchange.com/users/199893/bob-d), over at Physics.StackExchange, has already admitted to this NOT being a law so much as it is a convenience ...
        1. "Well, one can solve real life electrical circuit problems with a fairly high degree of accuracy using node analysis assuming no mass or material existence." – BobD Dec 7 at 23:16 (http://vinyasi.info/energy/stackexchange/Kichhof f's%20Laws%20vs%20Conservation.jpeg)
3. Assume from this act of *Conservation of Current,* create another sham and call it *reactive power* predicated upon the silly notion, not a fact, that reactive components store *reactive power.*
    1. This is a very interesting misrepresentation since *reactive power* is not power. Hence, it cannot exist in the real world. Instead, it disappears into the unreal world of imaginary numbers and remains there until it's time for it to come out, by way of conversion, into the physical world of real numbers and real power.
    2. In that unreal world of imaginary numbers, *reactive power* cannot interact with anything since it doesn't exist except within our imagination: the imagination which we ascribe to the square roots of negative numbers. The square roots of imaginary numbers cannot be solved in the real world outside of the make-believe world of our imagination bereft of logic. Logic is the cornerstone of mathematics. All of our mathematical operators can be logically deduced using Boolean Algebra forming the cornerstone of operations within our computers.
    3. Hence, *reactive power* retains its status of non-change. Hence, we misrepresent this status of non-change as *the definition of the storage of reactive power* since we tend to think of storage as a state of stasis as if energy found a way to "hide in a closet of non-changing status."
4. This makes the sham complete since we've managed to fool ourselves into believing that *reactive power* abides by the so-called Law of the Conservation of Energy and, thus, we have managed to seal up any hope for disputation of these illogical associations.
    1. This contradicts the fact that Mother Nature, in all of Her wisdom, created a material universe which abides by its Conservation of Energy. Then, She created a loophole through which energy could be *expanded* or *contracted* without any of it being *created or destroyed* and, thus, continue to support the Conservation of Energy outside of its reactance.
        1. *She* is not a hypocrite. *We* are the hypocrites since we want to believe in **a monopoly of Conservation**.
    2. She performs materialization in this manner because She wanted to be able to create new worlds **within the domain of empty space** without contradicting Her own Laws!
    3. So, forget about our misrepresentation of "perpetual motion machines" as running without any input of power, for *that is a lie.* They run on the expansion or contraction of electrical reactance after reactance has been converted from energy and placed into reactive storage as a temporary measure of isolation from the Conservation of Energy existing outside of reactive components.

I hope nobody gets burned at the stake for a fiction! (https://vinyasi.podbean.com/e/i-hope-nobody-gets-burned-at-the-stake-for-a-fiction/) – a 16 min., 20 sec. podcast on Podbean.

Cause and Effect

According to conventional streams of thought, there always has to be a "source" to supply a load. Thus, the source is considered to be a *causative agent* with consequential effects occurring at the load.

But what if this is not always the case? What if an overload of voltage at the load causes an escalation of reactance everywhere else within a circuit except at the voltage of the source?

Now, what can we conclude?

Possibly, that the source of voltage is not a source of energy so much as it is a source of stimulus much like the conductor of an orchestra does not have to play an instrument while he/she conducts everyone else to play theirs.

Nor does a band leader, or symphony conductor, have to feed his players fuel to sustain their actions since they may be performing for free out of gratitude towards the creation of their music and the beneficial outcome for their audience.

This is the way I view electrical reactance in which one segment of a circuit leads everything else without having to do too much of the work. Thus, an overunity circuit is often-times segregated into multiple sub-circuits each of whose section has a particular duty to perform: some sections will exhibit a lot of wattage while others will not. Some will be negative wattage while others will be positive wattage. Some will exhibit sine waves while others will show triangular waves or spikes.

Anomalous Kirchhoff Behavior

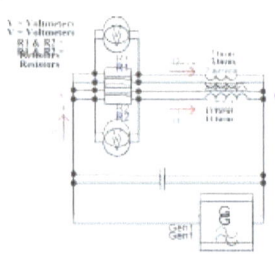

A plan is diagrammatically laid out of where, and how, to arrange the components of this experiment.

Splitting a transmission line into two branches should divide up the current, according to Kirchhoff's Laws, and maintain the same direction (ie, polarity) of current for both branches, yes?

This law, also called **Kirchhoff's first law**, or **Kirchhoff's junction rule**, states that, for any node (junction) in an electrical circuit, the sum of currents flowing into that node is equal to the sum of currents flowing out of that node; or equivalently: *The algebraic sum of currents in a network of conductors meeting at a point is zero.*

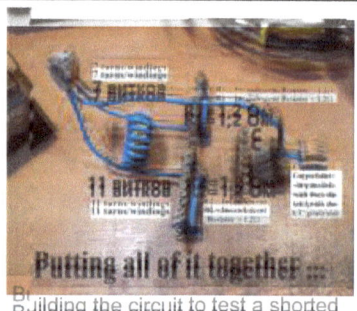

Building the circuit to test a shorted transformer.

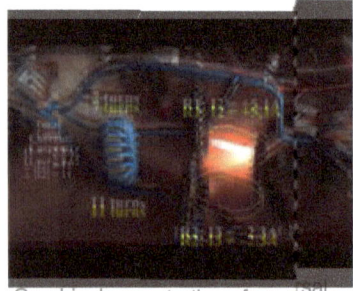

Graphic demonstration of reversal of current on the bench.

But what if conventional expectations are not always right? What if, sometimes, anomalous events can occur?

Take MrPreva's example (https://www.youtube.com/watch?v=XJnN3jk1Hyo) on YouTube translated by MrJohnK1 (https://www.youtube.com/watch?v=GFqJ5D0mkOo) and explained by Chris Sykes (https://www.youtube.com/watch?v=oXsXe9DJiXk) and others ...[88][89][90]

MrPreva has split a flow of current into two branches by shorting out both sides of a step-up/down transformer. Oddly enough, this shorted condition causes the current to reverse its polarity on the larger coil and add this negative current to the smaller coil which graphically heats up, and lights, his smaller coil into a luminescent orange glow! See, graphic demonstration on the right ...

Doing the math results in the following conclusion ...

Measuring the total amperage of the circuit: I1.

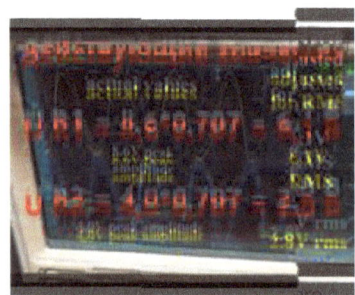

Calculating the root mean square (RMS) voltages.

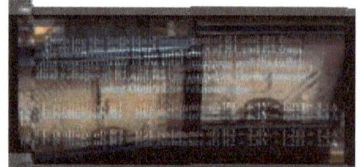

Calculations of amperage and voltage using Ohm's Law.

A fantastic conclusion, that: The Whole is Smaller than Some of its Parts

The current of I2 is greater than the total current of I1 due to the negation of I3.

```
I1 = 2.8A
I2 = 5.1A
```

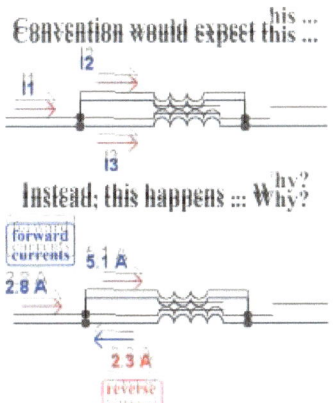

Convention would expect this...

Instead, this happens... Why?

forward currents
5.1 A
2.8 A
2.3 A
reverse current

Summation of results defying the initial assumption prior to performing this experiment:

```
I3 = 2.3A
```

Total Current of I1 = I2 + I3
2.8A ≡ 5.1A + (− 2.3A)
Yet, |I2| > I1 = *The absolute value of I2 is greater than I1!*

In other words, just because the whole is the sum of the parts *does not mean* that the whole is *greater than* any of its parts! Does this sound like a contradiction of the Conservation of Energy since Conservation dictates the form and function of Kirchhoff's Laws?

It is possible to simulate this, under ideal conditions, using Paul Falstad's simulator (http://falstad.com/circuit/circuitjs.html?ctz=CQAgjAzCAMB00JQVnBWSCCG1TADjwlQDYAmQlYKjacU286gUwFowwAoAN3DENWJ5e-UihjgOxWmGIZYAdmgIS8gCx5V8+WFXlwoRfMoRJoIgmHZDPX7BsceAE4OOrreYK96tDOeAMZ8LuAYlonKQrRwiVolam2uCiX7zBsw4WKfuYKnV9Y0CAOpLAeF+UB2AKqI+KyyWyZDxs4TyfL8L-J51Pho65yHu+pkt2J-7OHMH1EpdhAe08jC3yVTrdrE5gnP8B-tihJ1vNOXsshcZ5wjDoW9t4gAPLzeMjvMATjUAC4AeoCABoigA+CUckzFZMgBlMRMdq6f4gA8GtFAGEBG1JtAC-AAneoCAB0lgEc6gIAHSWAQDqAgAdJYBAOoCAB0lgEA6gIAHSWAQ1VgEA6gIFYv3wboDuA0cwDSBXv8GL3DOzyyVNkKIQxL3RBRBR4LZFHYvQxSVlFvjVpBBgGH3Bs5ANAjCtoS1rzOPQBBuCwv4oAKPw4AVqDogABEoAgpvzvF2+YgAlz4oAUuDJu+f4q0nxQAlcGpd+OC4AAKoAJRQA=) (the "Current inversion" image in the upper left corner) demonstrating that most of the A/C voltage source shifts into the realm of negative watts *all the time* rather than alternately every half-cycle!

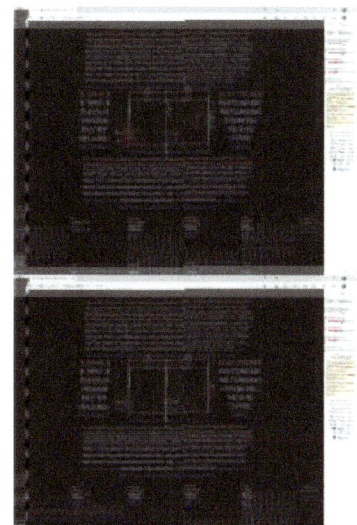

Fig. 9 − Reduce your energy expenditure, and lengthen the ride in your electric car (between taking pit-stops to recharge its batteries), by doing what all manufacturers of electric motors already know: add capacitance in parallel with inductance (http://vinyasi.info/ne?start&Circuit=powerfactor2.txt). It's so simple!

The use of capacitance, in parallel with an inductive load in Fig. 9 to the right, reduces energy expenditure since electrical reactance is recycled. This is in defiance to the electric company who will ignore this savings and charge us the same on our electric bill as if *conservation* didn't matter!

But this only works for inductive loads, such as: the electric motors inside of our electric cars; and the compressor inside of our refrigerator. What about non-inductive loads, such as: light bulbs?

They're taken care of by the insight of MrPreva (aka, "Pavel"), in Fig. 10 to the left, who has discovered that a shorted transformer whose two coils are of different inductances, will perform the same operation as the presence of a parallel capacitor will perform alongside an inductive load in Fig. 9 to the right.

By the way, ...

The capacitor, in Fig. 10, could probably be removed and the same benefits would occur. So, capacitance is not needed to perform energy conservation via its reuse within Pavel's example. All that is needed is a shorted step-up transformer and some resistive loads connected in series with both sides of this type of transformer. This is shown in the diagram of Fig. 10 in which two resistors are placed immediately above the transformer. One is labeled: 2.5 to signify 2½ Ohms while the other resistor to its immediate right is labeled: 7.5 to signify 7½ Ohms.

Application of Pavel's Discovery

is.gd/pavelsdisc

Pavel's shorted transformer, in Fig. 11a (on the left), can be used as a device for gaining overunity if we can figure out how to physically manifest a mutual inductance (a magnetic coupling coefficient) which is greater than one. Personally, I like to imagine that this condition represents a very large, additional mass of ferromagnetizable material = lying outside the transformer's core, yet = magnetically coupled to it. If this were true, then this would vindicate William Lyne's alledged quote of Nikola Tesla, when he said that: "... for every two hundred pounds of iron added to his Special Generator, (http://vinyasi.info/circuitjs/texts/Nikola%20Tesla/The%20Inventions,%20Researches%20and%20Writings%20of%20Nikola%20Tesla,%20e

Fig. 10 – Current inversion (http://falstad.com/circuit/circuitjs.html?ctz=CQAgjAzCAMB00IQVhBWSCcGITADjwiQDYAmQjYkJacU286gUwFowwAoAN3DFNWJ5e-UiRjgQxWmGIZYAdmgIS8gCx5V8+WFXiw6RfMoRSxI+spU4SDgHdhAoaVW0IgmHZDPX7iBsceAE4OOrre4Kq6tDQcAMZeLuAYIonkQrRwiAjYOFAu8GBI8qTs0HhIOPJW8NCcAM4hkQ6iVuIAZgCGADZ1TBwN4WkJPunt3b0cAObDAeF+UR4AKqj+KvyqJqj0erDEe-IwsLIQGYdYWJ7zXs4rYcWonkh817pPIvcQI-5uQhv8P0EQH8IrpgaQbrR5OgOMErqFbi9xKRofY4U1gfDoJ4wfc3l57lj7MCAcTtoSEQCriYMp5wvC0VEOAAPLzEMIYMArTmkDCg8BCADiTAAdkxAh0AC4Ae0CAB06gA+JXMoEQA48gBIMQ8qi6fAgAA8RvIAGEpcK6gBXAC2YpVOIcOHV+twtD1-CNBvIQtF4ulgXtaq8ZS88YGdnPdlAAgl0JWLhZKmHV5R1k3UOvakGENkkMAIdXQQFwxVa0xmWZAnKZQ1RwXnlz6xZKZfKIQrA+qyDXWShl2aLTaxa3IfY6U06xZKZfKIQrA+qyDXWShl2aLTaxa3IfY6U06xZKZfK...) can be simply achieved with a shorted transformer.

h.%2063.pdf) its output increased by one additional horsepower."[48] This extra iron was made available within the hull of a *very few* Elektroboots (electro-U-Boats (https://www.uboat.net/technical/electroboats.htm)) of Nazi Germany during WWII. William Lynes alleges that Tesla's Special Generator was bolted to the floor of the vessel's battery room insuring a firm magnetic coupling between his device and the multi-ton vessel which it was powering giving it a *minimum range* of 30k miles between recharging its batteries.[48]

By the way, ...

None of these specially designed Elektroboots were recovered after the war. Any of these vessels, which remained fully operational, escaped capture by the Allies. All we have are quotations from William Lyne, who is quoting a Mr. Dort, Jr., who (in turn) was quoting his father – Dort Sr.[48] For, it was Dort, Sr., who helped the Nazis adapt Tesla's Special Generator for use within a few of their Elektroboots. Some of these submarines were used to tow missile launchers carrying neutron bombs. These neutron bombs were another invention of Nikola Tesla, in addition to his Special Generator, and the liquefaction of air used to power these Elektroboots. The neutron bombs were tested in the Libyan desert by Rommel under the ruse of "looking for petroleum deposits." These inventions, and several more, were stolen from Tesla's lab in 1895. The fire, of this lab, (https://369news.net/2018/04/01/what-caused-nikola-teslas-1895-lab-fire/) was set to hide this theft. Immediately afterward, Carl von Linde of Germany patented Tesla's liquefaction of air giving a probable motive for this theft and arson (according to W. Lyne).[48]

On the other hand, Pavel's shorted transformer, in Fig. 11b (on the right), can *also* be used as a device for gaining overunity if its transformer wire gauge is 19 AWG or less and possesses a mutual coupling coefficient greater than 52.08%. This version is a lot easier to build!

Why Do I Like Pavel's Experiment So Much?

The reason why I took such a liking to Pavel's experiment is because it reminded me of something I did four years ago ...

I had just finished simulating the Joseph Newman device (https://josephnewman.info/) and had successfully demonstrated (five years ago) that he lied about how to build it. The version which he used for his demonstrations was different than the model in his book. The model in his book was an earlier version which never achieved overunity. Yet, the model which he used for his demonstrations *always* achieved overunity with the help of an engineer, by the name of: Byron Brubaker,[53] who told Newman to take out the rotating permanent magnets and replace them with dielectric canisters made from PVC sewage pipes, wrapped with an open coil, capped at each end with sewage pipe endcaps, and filled with helium gas. If you look at this picture,[80] you'll see that his so-called permanent magnets are wrapped with tape and painted to hide what they are. He never let anyone initiate his device. Only *he* was allowed to give several turns to his rotor while at the same time putting on a fake display that it was taking all of his muscles built-up from years of body building to get those so-called magnets up to the fastest speed possible.

The "Newman" secret was the fact that his main large coil was putting out a huge electrostatic field while the open coil surrounding his helium canisters was collecting some of it and translating their collection (of this electrostatic charge) to the helium. Once the helium became excited, it put out an electromagnetic field of its own at a much higher frequency than the rate at which these canisters of helium were rotating – as much as a million times per second of vibratory excitation (according to my simulations in LTSPICE).[91] This became a very large contributing factor for electrical reactance to boost the feeble current of his massive coil, which was wound with very thin wire. The massive coil doubled its input voltage which it was receiving from a pack of dozens of dry-cell batteries contributing their combined voltage of 350V to the coil's massively lengthy winding producing a 600V electrostatic field. Meanwhile, the contribution of current emanating from the helium canisters managed to reverse direction at a rate just under one ampere and head back to his battery pack to recharge them despite they were never intended by their manufacturers to be subjected to this form of torture! It's no wonder they wore out so fast! Newman refused to take Byron's advice and replace them with solar panels.

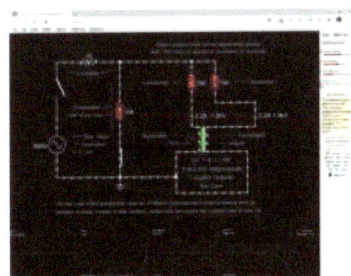

Fig. 11a – Pavel's generator (http://vinyasi.info/ne?startCircuit=generator.txt) does not need a prime mover acting as a source of power to maintain its generation of electricity. It merely needs a momentary burst from a sine wave generator of 1μV, plus a mutual inductance greater than unity, to initiate escalation towards infinite gain.

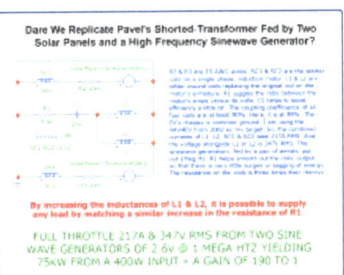

Fig. 12a – Pavel's shorted transformer is developed, here, into an overunity generation of voltage putting out 190 times more watts than it takes to run it – due to its frequency, various impedances located in appropriate places, its shorted condition, and its strong magnetic coupling. It is induced to draw lots of current from its sine wave generator despite its very low voltage.

I let this simulation sit for a year before I began to look it over to see if I could make any improvements to Newman's design, for I figured that Newman did not have the advantage of being educated as an electrical engineer (nor do I), and he didn't make use of simulations to help him design his products. So, why don't I give it a try?

What you see before you, in Fig. 12a and Fig. 12b and Fig. 12c, is the result of a few of my changes (to Pavel's design) which I serendipitously made four years ago. It's a shorted transformer[92] [93] just like Pavel's shorted condition of a set of parallel coils!

Thank you, Pavel, for inspiring me to remember what I had done four years ago but have long since forgotten! I am very grateful to you.

Fig. 12a is the schematic. Fig. 12b is its output. This simulation is designed to run within Micro-Cap 12 on a Windows 10 computer possessing a 64-bit architecture within its CPU making it impossible for round-off error to occur within its floating point, numerical notations. Fig. 12c envisions some hints derived from studying a single-phase A/C motor which I disassembled from a commonplace kitchen appliance: an ice-cream maker.

The schematic assumes a very tight coupling among all four coils at a minimum of 90%. This can be achieved by using ferromagnetic wire, such as: enameled iron wire used for floral designs, wound upon an iron core which is also capable of being used within PMH experiments popular on YouTube.[94] These experiments utilize the magnetic property of remanence exhibiting memory of the application of a D/C field resulting from coils which are wrapped around a toroidal iron core once their electrical excitation is turned OFF: they still continue to retain this memory for many years afterward. This is why this property was also taken advantage of during the two decades spanning the years from 1955 and 1975 within computer core memory before a better method was discovered.

It may also help to wrap these coils with aluminum foil to reflect their magnetic field back into their coils to help concentrate the intensity of their magnetism inside their core material? And maybe two turns of very stout, copper, solid core, insulated wire, wrapped on top of the iron windings and underneath the aluminum foil may also help? I don't know ... I'm just guessing.

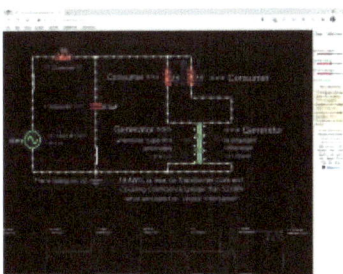

Fig. 11b – This variation (http://vinyasi.info/ne?startCircuit=generator2.txt) of Pavel's generator (also) does not need any prime mover acting as a source of power to maintain its generation of electricity, because a momentary burst from a sine wave generator of 1μV is sufficient enough to initiate its escalation towards infinite oblivion! But unlike Fig. 11a, to the left, this version uses a transformer wire gauge of 19 AWG or less and a mutual coupling coefficient greater than 52.08% to achieve its gainful overunity.

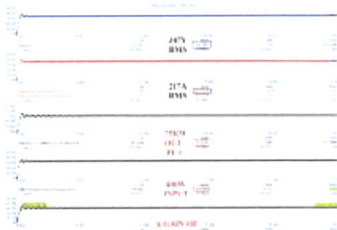

Fig. 12b – Here is the output for a derivative of Pavel's shorted transformer schematically displayed by its associated screenshot.

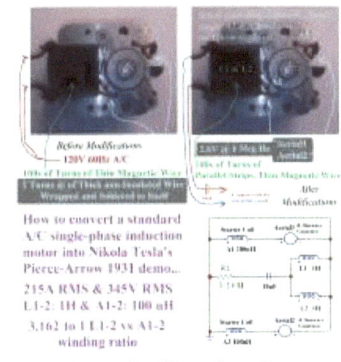

Fig. 12c – We could use some pointers on how to go about retrofitting a standard, single-phase, induction motor to make it behave in an overunity fashion. Here are some hints in conjunction with its updated schematic.

Revival of Archival Links

Inversion of Voltage Source for a Single-Phase Induction Motor (https://josephnewman.info/blog/f/inversion-of-voltage-source-for-a-single-phase-induction-motor) – Newman Motor blog

How to Make a Free Energy Device Which Clones More Electromagnetic Waves. (https://www.instructables.com/How-to-Make-a-Free-Energy-Device-Which-Clones-More/#step32) Step 32: Latest Discovery – Instructables.com > Circuits > Electronics

Inverted Placement of Source Voltage in a Single Phase Induction Motor (https://vimeo.com/vinyasi/invertedinductionmotor) – a Vimeo video

Fig: 13a – Pavel's shorted transformer is developed, here (https://ufile.io/is1fztk1), into an overunity motor putting out 320 times more electrical energy than it takes to run it probably due to its resonant impedance (https://www.translatorscafe.com/unit-converter/en-US/calculator/series-rlc-impedance/) and shorted condition:

Wikimedia Commons: Motor transformer (https://web.archive.org/web/20181210015332/https://commons.wikimedia.org/wiki/Motor_transformer) – a page at archive.org

Category:Motor–transformer (https://web.archive.org/web/20181210024110/https://commons.wikimedia.org/wiki/Category:Motor%E2%80%93transformer) contains the page, above, at archive.org

Download a compressed ZIP file (https://ufile.io/ubn2lb3b) of a Micro-Cap simulation of a variation of Pavel's Shorted Transformer/Generator – archived, Nov. 2022, at ufile.io

https://is.gd/downloadpavel (http://vinyasi.info/mhoslaw/Parametric%20Transformers/2022/Nov/13Spice%20-%20Tesla's%20Pierce-Arrow/Inverted%20EV%20Motor%20-%20Micro%20Cap/?C=M;O=D) – download simulations of Pavel's shorted transformer

https://is.gd/playpavel (https://www.youtube.com/playlist?list=PLwteYdUouDFCg3VTlls0-_HXxHSBJbFBQ) is a list of YouTube *free energy* slideshows derived from pop tunes

Fig: 13b – Here is the output for a derivative of Pavel's shorted transformer schematically displayed by its associated screenshot:

Using the Series RLC Circuit Impedance Calculator, (https://www.translatorscafe.com/unit-converter/en-US/calculator/series-rlc-impedance/) it is possible to improve performance of this derivative of Pavel's shorted transformer so much so that its operational frequency can be reduced to the point of it coming within the range of a normal rotary speed of a motor shaft while it is accelerating at maximum throttle speed. To calculate this, I used information which is provided by Darell Dickey on his website[95] concerning the first generation RAV4EV. At the same time, efficiency automatically improved bringing its coefficient of performance up to over 300 times more energy output than is inputted to run this simulation of a shorted transformer. This improvement adds the additional qualification of becoming a shorted motor.

It is significant to note that increasing the demands made upon the inductive load at Fig. 13a by increasing its inductance and series resistance to maintain consistency with its wire gauge size (estimated to be around 30 AWG) actually *lowers* how much power is drawn from the voltage sources labeled: *SolarPanel+SineWaveGen1* and *SolarPanel+SineWaveGen2*. This can be seen by comparing Fig. 12a with Fig. 13a and by comparing Fig. 12b with Fig. 13b. Meanwhile, the output goes up to match the load's increased demand! This results in a higher coefficient of performance as if to suggest that this design archetype rewards us for the demands which we make upon it rather than penalizing us with reduced performance or an increase of cost or both: *What a trip!* ;-)

This odd behavior (by conventional standards of expectation) is not so outlandish if we consider that this motor design is acting as its own generator while acting as a motor (at the same time) provided that its various parameters are met.

One very important parameter is that its two starter coils (which you may call: *current coils* since they are dominated by harboring mostly current along with very little voltage) must, each, be reduced to an extremely low inductance. In these series of simulations, they are both: 100 nano Henrys. This extreme smallness is not unusual since these are intended to be equivalent to the self-shorted, starter coils of standard, single-phase, induction motors utilizing very stout, non-insulated, self-shorted, twice-wound copper wire which is usually wound through holes bored into the motor's armature as well as tightly wound around the armature of these types of motors as can be seen in the inset-image of Fig. 12c (labeled: *motor hints*).

This style of motor is able to become its own generator due to the reversal of voltage, or of current (but not both). Pavel's demonstration of the reversal of current has some peculiar results (as can be seen in the figure to the right). In *his* case, the reversal of current on his larger coil failed to send its current back to its source of his A/C power. Instead, it succeeded at dumping all of its current onto its adjacent coil causing that coil of lesser windings (whose lesser inductance could not resist this contribution of extra (current) to increase its own current beyond the level of current which is feeding both of these separate branches. Please see the simple math, above, labeled: **A fantastic conclusion, that: The Whole is Smaller than Some of its Parts**:

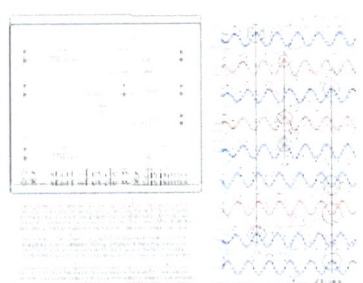

Fig. 14 – a segregated analysis (https://www.bing.com/search?q=The+Meaning+of+Unity+in+Energy+Conversion+Systems&FORM=SSQUIC&PC=U531&lightschemeovr=1) of each electrical component within Pavel's generator in an attempt to discover if there is any inversion of voltage or current which may indicate the generation of *free voltage* or *free current*. If any of these inversions are found, then this means that the electrical components which are hosting these inversions are no longer passive components (according to passive sign convention), but have become *active components* without assistance from *any* prime mover.

By definition, a prime mover is an energy source which *has to* inject its energy into a device from *outside* of that device. This conventional requirement, which we make upon *all of our appliances*, makes it possible for us to deny whether the colloquialism known as *free energy* and *overunity* actually exists. But as you can see, at least in theory, *Conservation of Energy* is restricted in its scope applying exclusively to situations that require a prime mover to supply all of a device's power **without** *any generation of additional power arising from the device, itself*. This, of course, is **a rule of thumb = not the** rule of Law, since amplifiers are well-known instruments of magnification.

Herein, simulation Fig. 14 (https://ufile.io/59dxctya), is no exception. Amplifiers are by no means non-existent. To insist that amplifiers also be under-achievers is nothing less than a straitjacket of conformist domination made upon *free thought* and *free action among free people*.

An afterthought ...

Any quantity of change to both input voltages of each SolarPanel+SinewaveGen1+2 in any of the circuit simulations of Fig. 12a, Fig. 12b, Fig. 13a, Fig. 13b and Fig. 14 results in the squaring of their results. This is in conformity to the square=cube law and suggests that *energy* is a manifestation of the surface area of some object or phenomenon or both? So, if both input voltages are increased by a factor of three, then their resulting outputs will be increased by a factor of nine. Or, if their input voltages are decreased by a factor of one-third, then the outputs of their circuits will decrease by a factor of one-ninth.

Of course, solar panels could be replaced by batteries since the input voltage is merely 1½ volts for each sine wave generator, voltage source. And each of these two "sources" are putting out a frequency of 190Hz. The batteries can be swapped out and replaced with freshly recharged batteries while the spent batteries will be getting themselves recharged using a tiny portion of the output of this scheme. So long as the batteries don't wear out, this scheme will provide unlimited energy from nowhere existing outside of this setup, but coming from within this device itself!

Shorting a Transformer Divides its Current

This is (http://vinyasi.info/realsim?startCircuit=current-division.txt) what Pavel's discovery manages to accomplish: the higher-resistance, higher-impedance side of his transformer shunts its current over to the lower-resistance, lower-impedance side. That's why it has so much current. Yet, this more conductive side has so much current since the other side of the transformer won't accept very much current. So, most of it has to go to the side of lesser impedance and lesser resistance.

Mutual Inductance is the A Priori of Negative Resistance

Up until writing this section, I thought that reversal of current was the *a priori*, first cause[36] of negative resistance. I'm wrong. The non-saturation of mutual inductance is the first cause while the reversal of current is its diagnostic check and negative resistance is the label which we pin to all of this. We can't see the non-saturation of mutual inductances, but we can infer that this occurs among the reversal of currents of its participating self-inductances. This non-saturation of mutual inductance causes an inversion of the polarity of the magnetic fields surrounding the participating self-inductances. This, then, results in their reversal of currents which we label as being:

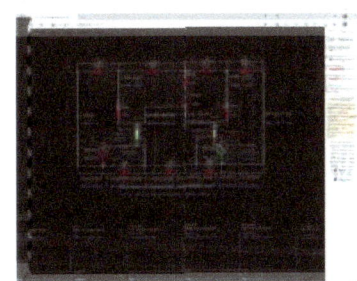

Current division via a shorted pair of transformers.

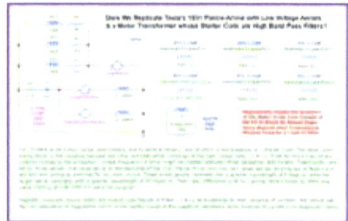

Fig. 15a – This simulation demonstrates the *a priori* of negative resistance lies within the magnetic field of mutual inductance.

Fig. 15c is a closeup view of the output for schematic Fig. 15a exhibiting no indications of non-saturated, triangular waves within its self-inductances.

negative resistance – *although negative impedance would be more accurate*. In fact, we have upgraded our nomenclature to become: *negative differential resistance* (https://www.bing.com/search?q=negative+differential+resistance&FORM=SSQUIC&PC=U531&lightschemeovr=1).

The addition of an extra winding (called: Squirrel Cage Rotor), which is electrically isolated from the rest of the circuit in schematic Fig. 15a, makes plainly visible that the inversion of currents among all of its self-inductances were occurring all along within the prior stages of the development of this archetypal circuit demonstrated in Fig.s 12, 13 and 14, up-above. We merely couldn't see these inversions of magnetic and current polarities, but they were there hidden in not-so-plain sight.

As is evident in Fig. 15b, the capacitor is a consumer of electric power since its surging voltage and amperage tracings are both headed in the same direction of polarity, namely: they're headed in a downward direction of increasingly enlarged negative amplitudes by comparison to the vicinity of the zero midline from where these tracings began.

All of the other surges are split between the polarities of their voltages and their currents, each headed in opposing directions of polarization, indicating that all six inductors have become generators of negative watts. This behavior is replicated within all three sine wave, voltage sources. There is hardly anything within this circuit which does not generate negative watts. It's no wonder it wants to surge to infinite gain!

Fig. 15c gives no indication of any initiation of triangular waves riding on top of sine waves as was apparent in the Growth of Triangular Waves in the so-named section, below.

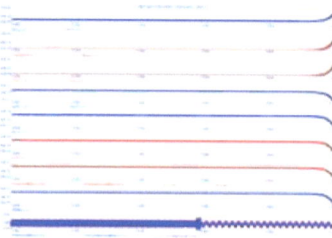

Fig. 15b is the output for the schematic of Fig. 15a.

This circuit is not intended to be built. Like the circuit, below, within the section which is labeled: Non-Stable Output of Extreme Overunity, Fig.s 16 and this circuit, Fig.s 15, are merely useful for demonstrating (by way of inferential hint) where lies the first cause of the growth of overunity.

Non-Stable Output of Extreme Overunity

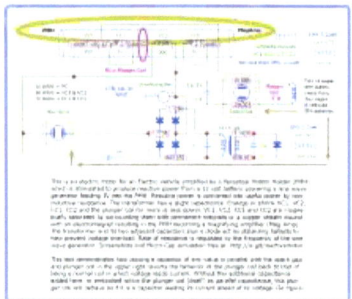

Fig. 16a – A greater quantity of nodal voltage, for this circuit simulation (https://ufile.io/qiar4tku), accumulates on the side of the smaller sets of coils (of this lumped set of inductors) than on the larger side in deference to the behavior of a normal transformer which has not been shorted between its two sets of coils.

I always thought it strange that Pavel's experiment accumulates a preponderance of voltage among the smaller coils of a shorted transformer in deference to the conventional behavior of a non-shorted transformer in which a preponderance of voltage always accumulates on the larger set of coils. But now, the nodal voltages in this experiment, of: Fig. 16a, (https://ufile.io/ze260928) accumulate a tremendous quantity of voltage on the smaller coils, labeled: CC1 and CC2, just like in Pavel's experiment. *Go, figure!* If anyone can figure this out, I'd like to know about it. Just send me an explanation on one of my user: talk pages. Thanks!

Fig. 16b shows how stable is the output of this type of circuit in which it does not explode, within its four second duration of run-time, with infinite power due to the periodic collapse of its triangular waves (exhibited in extreme closeup in Fig. 16c) which prevents the possibility of any out-of-control explosion *at least for a little while!* But wait until after four seconds. You're in for a treat. This circuit blowups like crazy!

Removing the spark gap at the rotor coil stabilizes it and prevents any possibility of explosive escalation of the amplification of amplitude up until 16 seconds of run-time. Further than that, I do not know since my computer doesn't have the RAM to handle that much duration of simulation. But, this variation of Fig.s 16a+b+c forms the basis for my slideshow of 15 images depicting overunity riding piggy-back on top of Thermodynamic Equilibrium (see the following sections entitled: *Times have Changed* and *Growth of Triangular Waves*.

But, how do I know that a physical build of this type of circuit (https://ufile.io/jcsq1fvz) won't want to initiate sparking? I don't. It has such high nodal voltages that it's probably going to turn ON a plasma state of arcing which will definitely push the output of this circuit into a sudden and complete explosive destruction of itself. You can look it up on my website under this directory: Inverted EV Motor - Micro Cap (http://vinyasi.info/mhoslaw/Parametric%20Transformers/2022/Nov/LTSpice%2

0-%20Tesla's%20Pierce-Arrow/Inverted%20EV%20Motor%20-%20Micro%20Cap/?C=M;O=D) under the name of: reactive-motor-v3b.zip. (http://vinyasi.info/mhoslaw/Parametric%20Transformers/2022/Nov/LTSpice%20-%20Tesla's%20Pierce-Arrow/Inverted%20EV%20Motor%20-%20Micro%20Cap/reactive-motor-v3b.zip)

The triangular waves of these inductor's voltages, of either of the large coils which are labeled: VC1 or VC2 in Fig. 16c, indicates that these coils are not saturated and cannot reach saturation of their voltage or their current. They *must* allow for an infinite escalation of amplitude if this circuit were to be slightly modified to prevent its periodic stable collapse of its amplitude and encourage, instead, a highly unstable rise towards the infinite oblivion of this circuit's existence if it were to be built.

Times have Changed

The science of electrodynamics hasn't changed. Its focus has changed.

A hundred years ago, it was possible to pick up a book[97] which emphasized the differences between positive and negative resistances within an oscillating circuit involving both "forced oscillations" impressing themselves upon a circuit by entering into it as its input source of EMF and then resulting in the formation of "free oscillations" acting as the output of that type of circuit.

But times have altered the focus of electrical engineering to the extent that overunity is not discussed anymore. It's not that it was never discussed. It's just that no one is old enough, or still alive, who remembers it being discussed let alone taught to budding students of electrical engineering.

Good ideas never die. But they do need to be revived from time to time. Especially if these good ideas are timeless classics.

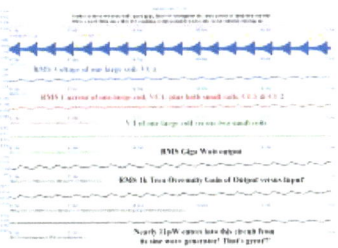

Fig. 16b – Virtual oscilloscope tracings of Fig. 16a displaying nearly 11μW of input power which could be garnered from less than half of the output (https://electricityandalternativeenergy.quora.com/How-many-watts-can-I-expect-to-get-from-an-array-of-micro-mini-solar-panels-putting-out-2-6-V-2?ch=10&oid=401787542&share=daa129a4&srid=3zXXZ&target_type=answer) of a very small single solar panel (https://www.mouser.com/ProductDetail/PowerFilm/INP3.6-12x310?qs=vvQtp7zwQdP%252BiHu4s5Cyzw%3D%3D) (archived (https://web.archive.org/web/20221204173320/https://www.mouser.com/ProductDetail/PowerFilm/INP3.6-12x310?qs=vvQtp7zwQdP%252BiHu4s5Cyzw%3D%3D)).

Growth of Triangular Waves

is.gd/negdampslides

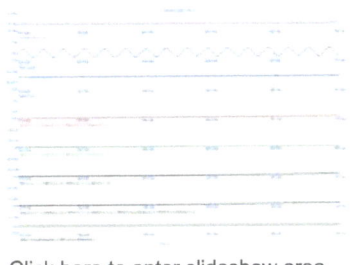

Click here to enter slideshow area.

Growth of triangular waves ride piggy-back upon the input of sine waves whose non-variant amplitude are ever-present within the voltage tracing of the large coil, VC1, within the figures which are contained within Category:Growth of electrical non-saturation.

HINT
Play all of the slides within Category:Growth of electrical non-saturation by hovering your cursor over the dark green square on the far right. It looks like this: > at first. But after you hover over it, it will change into this: Show Slideshow > and then you'll be able to click on it to play the slides.

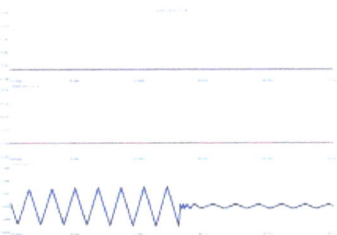

Fig. 16c – Extreme closeup view of triangular waves of voltage on one of the larger sets of coils of Fig. 16a indicating its non-saturation of voltage. This implies that its current will fail to saturate as well as its voltage. This makes it possible to achieve infinite growth of amplitude should these triangular wave-forms persist throughout this circuit's operation.

The sine wave input, which underlies these triangular waves, retains the same frequency and amplitude over time. The sine-shaped, carrier wave represents the "forced oscillations" of voltage input entering into this type of circuit from its sine wave generator. This input source of voltage provides the circuit (to which it is attached) only one terminal (half a network portal) for current to escape, namely: its terminal of entry. Hence, its coils and capacitors are *forced to generate* free oscillations of non-saturable, triangular waves of an ever-escalating amplitude (of both voltage and current) in order to free itself of its confinement from within its half-portal network. This is the only way in which voltage buildup may escape confinement: is by reversing its polarity of direction relative to voltage and buildup to the point of explosively exiting the circuit by destroying its host-circuit.

Convention dictates that we provide current with a throughput to encourage its natural inclination to flow. But my unconventional approach is intended to thwart that outcome and deny a throughput of current since that would result in a relative synchronicity of positive unity, power factor, and a boring output less than its input satisfying conventional expectations of under-unity. We want a negative unity, power factor or else the generation of free energy will not happen. So, we have to risk destroying the circuit (in pursuit of free energy) if we should fail at regulating its explosive growth of overunity. But that's the price I am willing to risk if I want free energy to manifest.

So, we could safely deduce that the forced oscillations have already saturated this type of circuit configuration while the free oscillations will *never* saturate this circuit.

It's as if this circuit is exhibiting the characteristic behavior of a multiplex network which can accommodate multiple streams of data transfer due to the unique characteristics of each data stream: the forced oscillations remain saturated while the free oscillations do not.

Hence, the amplitudes of voltage and current of the forced oscillations remain constant exhibiting their Conservation of Energy while the amplitudes of voltage and current of the free oscillations continue to grow beyond the RMS amplitudes of the forced oscillations. This would violate the Conservation of Energy if these free oscillations were to be considered as the resulting output of their causative input for that would seem to be the case at first glance.

But first impressions can be deceiving.

If I vary the input voltage of the sine wave generator, not much change occurs at the output. Although some direct relationship exists between the two, it's not an equivalency since the input varies by a fraction (and not linear) to the variation of the output. It's more like an exponential differential (at the very least) barring explosive rates of growth transforming their exponential relationship into an infinite rate of growth.

If it be true that Conservation of Energy is a Universal property of all forms of energy and inviolate, then I am left to conclude that reactance is a self-feeding process which contributes to the overall amplitudes of energy.

In other words, ...

Capacitive inputs (the preexisting conditions of various capacitances within a reactive circuit) become the outputs of capacitive reactance which, in turn, become the inputs of capacitance for the subsequent cycle, or half-cycle, of oscillations, and likewise for inductive inputs.

These capacitances and inductances grow over time and, thus, give the *appearance* of the violation of energy conservation without actually violating anything due to our erroneous presumptions are affecting what we conclude is happening.

But, when one capacitor can affect the capacitive field of another capacitor, and when one inductor can affect the inductive field of another inductor, and when two capacitors can affect the field of two other inductors, and vice versa, then free oscillations are free to expand their amplitude over time without anything to stop them (much less Conservation of Energy).

An electrostatic field surrounding a capacitor is the result of capacitive activity. And a magnetic field surrounding an inductor is the result of inductive activity. So, energy does play a role in the growth of overunity, but energy does not bear sole responsibility for that would violate the Conservation of Energy.

This is analogous to how some criminal minds think.

When one mafia boss wants to launder his profits, he channels them through another mafia boss, who channels them through a third, and a fourth, etc, to hide what each is doing and claim innocence. Spammers do the same thing. They relay their spam all across the globe, bouncing their emails against multiple servers until it becomes a blooming mess of complicated accountability.

Well, free energy is no exception! And free energy enthusiasts could be considered as outlaws based on how society judges the accountability of its members, but without digging any deeper (such as what the IRS always does or a private investigator).

Energy enters a reactive component within a circuit. This reactive component absorbs this energy and converts it into energy by first passing it through its dynamic field of reactance acting as an intermediate middle step of conversion (aka, laundering the cash). So, in reality, energy does not exit upon entering a circuit. That's only true for nodes which lie in between two components according to Kirchhoff's Laws. It does not hold true for the components, themselves. That's the catch.

Thus, Energy Conservation has nothing to do with the components of a circuit. And much less does Energy Conservation hold true for reactive components of a circuit. Energy Conservation only holds true for the conceptual void (which we call nodes) which lie *in between* the components of a circuit, which is like saying that Energy Conservation is limited, ie. restricted, to nothingness.

Conservation of Energy controls fluff.

Do you always pledge allegiance to fluff?

There's nothing wrong with the Conservation Law. It's the socially provocative, peer pressure to pledge allegiance to nothingness which disturbs me.

I remained silent, back in the days when we were supposed to vocalize our allegiance to the flag of our country.

What did you do? Blind obedience?

Free Energy Generators don't have to be Fake

Attempts at producing free energy don't have to be fake. There's no guarantee, either way, for them to be fake or real, since there's a general lack of knowledge on this complicated subject ...

Please see :::
Ohm's Law forms the foundation of basic electric theory in conjunction with the formulae of electrical reactance: (https://vinyasi.podbean.com/e/ohm-s-law-forms-the-foundation-of-basic-electric-theory-in-conjunction-with-the-formulae-of-electrical-reactance/), is a podcast at Podbean.

::: and :::
Is it possible to make a self-powered free energy generator? (https://www.quora.com/Is-it-possible-to-make-a-self-powered-free-energy-generator/answer/Vin-Yasi), is an answer to a question on Quora.

Free energy generators sometimes catalyze their reactances off of freely available voltage potentials within the atmosphere at ground level amounting to a few microvolts which were enough to run crystal radio sets a 100 years ago and are enough to stimulate the over-reactance of a circuit which is starved for power and possesses only one portal; an inlet, for current to flow. This prohibits the formation of normal current. Instead, current finds no other outlet than to reverse itself and exit the same way it came in. And since this is reversed current, relative to voltage, then this manages to increase voltage differences rather than equalizing them and is the definition for the generation of electric power under passive sign convention. This has the consequence of converting passive components, such as: coils of wire, into active components (which generate power rather than consuming them). This offsets the miniscule power entering this type of circuit and replaces the significance of prime movement since the activation of passive coils becomes its own significant prime mover. The catalysis which arises from the environmental background voltage remains significant, but not because it delivers power, but because it is a dependable resource to rely upon since "perpetual motion machines (who do not require any input of power)" are not a fantasy so much as they are a misrepresentation of the reality of free energy augmentation.

The generation of free energy is a multistep procedure involving the acquisition of nanowatts or picowatts of power from our environment to act as a stimulant to catalyze over-reactance of pairs of coils working in conjunction with pairs of capacitors, followed by the augmentation of this reactant voltage via an open transmission line archetype, followed by a self-shorted transmission line archetype (made manifest as a self-shorted coil) to offer a pathway for current to flow. This current will remain in opposition to its voltage and may not require its shift of phase (by 180°) to correct for this power factor since it is already usefully sending power back to its source!

Errors and Oversights I have discovered on the Internet

Frederick Alzofon's Unified Field Theory

1: "One fundamental feature of the unity of nature is that only two forms of the reality observed exist: radiation and matter." – David Alzofon quoting his father, Frederick, on page 160 of David's book on his father's theory of *Gravity Control with Present Technology* (https://a.co/d/gAcNw8p). This quotation is from Frederick's paper, entitled: *The Unity of Nature and the Search for a Unified Field Theory*, F.E. Alzofon, p. 600, *Physics Essays*, 6 (1993) 599-608. [98] [99] [100]

   1: I beg to differ ...
   2: My first response to Frederick Alzofon's Unified Field Theory, (https://vinyasi.podbean.com/e/my-first-response-to-frederick-alzofon-s-unified-field-theory/) a podcast of 33½ minutes.

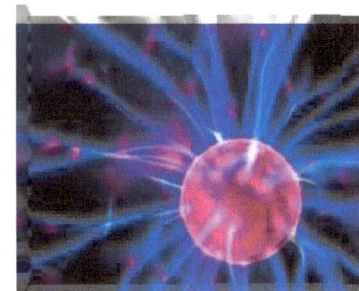

A plasma globe.

Bowning Intellectual Property, Ltd.; UK

I sent an email to this company after I discovered an oversight on their webpage devoted to helping people understand why it's so difficult to obtain a patent for overunity devices is often due to the assumption, by the patent office, that it won't work and (thus) will be useless. This will automatically fail one of their criteria for submission as a utility patent.

Subject: error on your webpage

You have an error on your webpage...
How to patent a perpetual motion machine (downing-ip.uk) (https://www.downing-ip.uk/blog/how-to-patent-a-perpetual-motion-machine)

For, it was 1921 and the Ammann brothers had just demonstrated their batteryless EV on the streets of downtown Denver with two newspapers there to document the occasion.

One of the two brothers failed to get a patent, not for lack of trying, but for getting arrested and charged with "stealing energy from the grid" when he drove his batteryless EV into the jurisdiction of Washington, D.C., on his way to the Patent Office. Sounds like the brothers were getting acknowledgment, in a roundabout sort of way, for successfully creating a device which required no batteries to run their EV?

Announcement of the Ammann Brothers' demonstration of an EV conversion without batteries preceding Tesla's Pierce-Arrow demonstration by ten years.

The error is that it is impossible to isolate systems of energy from their surroundings. Hence, perpetual motion machines are not possible, not due to the laws of physics, but because the laws of physics, which define Conservation of Energy and thermodynamics, are predicated upon a fictional assumption that – what was intended as a mere hypothesis for alleviating the difficulty of accounting for all of the energy of a system – has been converted into a Law as if it was also applicable to the physical world.

When I wrote my submission for a provisional patent, (http://vinyasi.info/mhoslaw/Provisional%20Patent%20Application%20-%2063221840.pdf) I took the position that energy accountability of every single component in an electrical system had to be accounted for – especially, if that device purported to be overunity. So, I did analysis on various devices, beginning with a simple simulation of a flashlight circuit, to assess whether or not energy output was less or greater than energy input and hopefully also be able to make a determination as to where did this extra energy come from or disappear to.

I failed to make that determination. Some circuits do appear to possess an output which is less or more than their input. But there are no clues as to where that extra energy disappears to or comes from.

It's no small wonder that physics turns a blind eye to this mystery.

But Washington, D.C., solved the problem by arresting C. Earl Ammann a century ago for stealing public property. Maybe that's the solution to this mystery of physics which has not ceased to intrigue people. There are many inventions on YouTube claiming to be overunity. Maybe they are overunity. And maybe it is their unique style of reactance which pumps energy into those devices from somewhere outside of themselves using reversal of current to accomplish their task of increasing their internal differences of voltage against the natural order of entropy?

Who knows?

A Few Criticisms

Quora Criticism #1

1. Here are a few of my comments within a post at Quora ...[101]

My response to a critic:

> "You cant boost nothing"
>
> I wouldn't consider you to be nothing. You are a living, breathing creature capable of running the electrical force in your body at a level of wattage little different than the background wattage of our environment. That is the amount of energy I prefer to feed the simulation of my circuits. True, sometimes slightly more, but no more than about 3V. But that's only under specific and rare circumstances. Usually, I like to keep the voltage down around a microvolt which amounts to nano watts or pico watts which is how much energy you and I run off of.

Laughing monkey.

This is also similar to how much power was supplied to crystal radio sets of 100 years ago. But the difference is that you and I are very efficient at how little electrical power we run off of by comparison to crystal radio sets from a bygone era which were far less efficient putting out, at best, a feeble signal to their earpiece.

You must've heard stories of grandmothers, in a fit of passion or whatever it was that overtook them at that moment in time, lifted up a car to get one of their grandchildren out from under it defying the laws of physics! How did they do that? Or is it just a lame attempt at wowing us with fantasies?

They probably spent a lifetime of devotion in a fraction of a second if I were to take a guess at how they *did not* violate any law of physics any more than spending a battery in a moment by discharging without any resistance would cause that battery to blow up!

It's true that energy is forever in a state of being frozen unless acted upon by forces exterior to itself. But that does not mean that electrical reactance is also a fixed condition incapable of flexibility, because electrical reactance is not referenced to its environment. Unlike energy, reactance references to itself. This is why reactance is so readily manipulatable.

And since energy and reactance can be converted from one into the other and back again, effectively speaking, energy can be magnified or contracted but in a roundabout fashion via its conversion into electrical reactance and the reactant manipulation which we can perform upon our perception of energy whenever it is being charged and discharged to, or from, its reactive containment.

Electrical reactance is more than just a temporary form of storage. It is more than this. It contributes the characteristic of its containment onto the energy stored (inside of it) altering the characteristic of that energy in ways that are determined by the potential characteristics of electrical reactance, namely the characteristic of: frequency, phase shift, capacitance, and inductance, all of which determine the final product of energy which discharges from the temporary phase of storage of reactive energy.

Take frequency, for instance ...
A second of a kilowatt is not the same as an hour of a kilowatt. And that's just one example of energy containment affecting how, or to what extent if any, the discharge of that energy is going to have upon its surrounding environment, namely: the other electrical components affiliated with its location within that circuit.

It's like a speed skater who gets a boost from a jet pack strapped to his body causing him/her to scoot far ahead of the other speed skaters. Because, if he/she were to time compress their skating to the finish line, it wouldn't matter how much energy, nor would it matter how little energy, they possessed. Nor would it matter how fast or slow they skated. They could be skating slower than everyone else, from their perspective. Yet, due to a contraction of time, they'll still win the race. That's what impact reactance can have upon the outcome. It's a delusion to think that we know how to measure the quantity of energy whenever reactance can override our measurements causing us to think that some law of physics has been violated, which we know cannot be, so we'll bury the data and move on to something else ignoring what our training has taught us all these years: that energy cannot be created nor destroyed. True, but largely not relevant under the circumstances which I am describing and exampling.

So, the physic's Conservation law, and the Kirchhoff Laws, which states that energy IN has to equal energy OUT only applies to real power. It does not apply to electrical reactance since there is no such thing as another variety of power called: reactive power. That would be a contradiction in terms allowing for the ability of imaginary numbers having any impact upon real numbers. They can't. That's why we keep them separate from each other and call this a: complex situation!

In other words, when energy enters into a condition of electrical reactance, what comes out ... the electrical reactance which comes out ... is not the same as the energy which entered into that reactant condition. It could be more reactance than the energy which entered, or it could be less, but it is never the same. Only energy IN equals energy OUT. Kirchhoff Laws do not designate anything other than the consistency of current and voltage. Kirchhoff Laws ignore reactance. So, there can be no equivalency between reactance IN versus reactance OUT. And if a law of mankind, a legalistic law, fails to prohibit a thing, then you can be sure that that thing is allowed by law.

It is a known precept of political law that if something, anything, is not explicitly stated somewhere within the labyrinth of laws, then it cannot be prohibited.

The inlet of reactance cannot equal the outlet of reactance because reactance is incapable of maintaining itself since reactance is a mechanism of self-reference unlike energy which always refers to something else other than itself (usually, outside of itself). Energy and reactance are completely different in how they go about "grounding" their reference point.

So, your concerns are valid from a simple viewpoint, such as that of a child, but I am addressing you as an adult knowing that you can handle something more complicated than what children are expected to handle.

Unfortunately, electrical engineering is so complicated that we are all children on some, or another, or all of the subtopics of electrical engineering. So, it is no surprise that the public is underinformed, and misinformed, riddled with con artists who take advantage of our pandemic ignorance.

Electrical engineering taught us the facts of its subject and gave us permission to use its most important feature for purposes outside of the limitations of energy if we ever bother to test conventional wisdom to see if it conforms with whatever we've been taught.

Unfortunately, I don't count a lecture as superseding experience. If lab experience should differ from the lecture, then either I don't understand what the teacher is driving at, or the teacher has overlooked the freedom which the experience of laboratory experimentation can offer to anyone who is open to it.

Instead, our formal training emphasizes the domain of real power (involving volts and amps) and the Kirchhoff Laws to back up that limited range of usage. We are never taught that the container of energy, namely: reactance, is more powerful than the energy which is laid out by Ohm's Law since reactance manipulates perspective. Only energy is actually "contained" by reactance. Reactance, itself, cannot be contained by anything since it supersedes the influence which energy may have.

It's true that reactance can electrocute living creatures. Energy does that; reactance cannot do that. Yet, reactance can alter energy so that its prior state of energy (which could not do any harm) could easily become transformed into harmful energy.

To finish off this comment, ...

Imagine, if you will, a king who cannot be a king if he has no subjects to be a king to rule over. And this king has no family to pass his legacy of rulership onto when he dies. This is the dilemma of reactance. If reactance has no energy to count as its subject, then nothing gets done. But once energy comes onto the scene, no matter how small and insignificant that energy may be, it doesn't matter. For, that king will treat that paltry energy as if it were his lost prodigal son giving this adopted son endless banquets in his son's honor. And that king will love and devote his remaining years to the welfare of that adopted son as if nothing else mattered. For, that king must have an heir to his thrown. Nothing else matters. He'll love that son, no matter what, even if that love should kill that king.

One more comment of mine from that Quora post ...[101]

When we manage to boost the energy from solar panels operating at night, such as in the manner of boosting input power which my scheme suggests at over 300 times magnification, then we'll have 24/7 power from solar cells.

Using the data from ...
Radiometry and photometry in astronomy (stjarnhimlen.se) (http://stjarnhimlen.se/comp/radfaq.html#10)
Full sunlight is 130k lux. Starlight with airglow is 2m lux. That's a ratio difference of 65 million to one. So, ...

The cube root of 65 million is approximately 400. And my electrical boosting scheme delivers a magnification of 320. So, my scheme is capable of boosting nightlight falling upon solar panels if more than three of these setups are daisy-chained so that each output feeds the input for the next module in series.

Quora Criticism #2

2: Here's another criticism, plus my response, to another posting on Quora ...[102]

My response to a critic:

"That is a complete load of rubbish, throwing away most of the solar panel voltage throws away most of the available power. Statements talking about current reversals are also quite useless to the job of producing energy, there is no free energy, it is at best an illusion, this author is a charlatan."

Is the author's use of English that confusing? I thought it was clearly about expanding reactance under temporary storage in the complex field in and around reactive components. I thought reactance was for free apart from the initial investment of the circuit's design and construction plus a catalyst style of prime mover, not a prime mover with the intention of providing any significant power.

A catalyst, by definition, is reusable. But a catalyst, by implication, can also be quantitatively insignificant yet qualitatively very significant enough to overcome its quantitative insignificance and make a valuable contribution towards efficiency.

A large part of reactive efficiency is all about its reusage per unit of time so that more reactance can be reused per unit of time rather than less reusage. This boosts coefficience of performance figures closer and closer to unity so that an electrical system which uses one unit of power at its input and yields 100 units of power at its output (all due to the reuse of reactance) will have an efficiency of 99%. That's not overunity of efficiency despite its output gives the appearance of being overunity when it is not, really, despite its appearance.

So, maybe the author is being purposefully, or accidentally, vague on what the term of overunity is being attached to: efficiency or output, since it clearly only applies to the latter. But since this is a situation of the reuse of reactance, overunity is not very important – nor is it very significant – since it's not at all accurate, because there is no overunity of power. It merely looks that way at first glance.

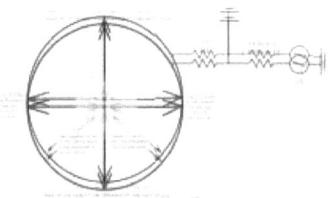

Oscillations of Radiant Energy due to throwing away most of the input and prohibiting the formation of current (within this circuit) by disallowing an exit (to avoid satisfying its inlet). For neophyte designers of overunity circuits, there should be only one inlet doubling as its own outlet.

Facebook Criticism #1

1. Here's a criticism, plus my response, to a comment from a friend on Facebook …[103]

My response to a friendly critic:

"I heard otherwise and it's by a TM Governor (https://governors.tm.org/)!"

I quote the editors over at Wikipedia who contributed to the article entitled, isolated system. That's one of my points. I also cite some of my simulations that indicate that it is not that hard for a so-called free energy circuit to steal energy from a transmission line. And since it's impossible to conduct a controlled experiment to rule out this type of theft, and since the utility grid absorbs energy from its surroundings just like a free energy circuit does, then it is impossible to assume that free energy comes out of nowhere or that perpetual motions machines exist (since our definition of perpetual motion machines is predicated on the assumption that they are isolated energy systems: "A perpetual motion machine is a hypothetical machine that can do work infinitely without an external energy source." – Perpetual motion). So, if I'm wrong, then my references are wrong as well, for I did not take them out of context. In fact, I doubt the universe is an isolated system, for no doubt, according to Krishna who is quoted in the Bhagavad Gita as proclaiming that Arjuna should get up off of his sorry butt and fight. For, if Krishna were to rest for a single moment, then the whole of creation would come to a dead stop.[104] Creation is predicated upon the constant dynamic quality of Krishna. So, if that is not an admission of the universe being a non-isolated system of energy, then I don't know what is. Every particle in Creation is in constant communication with all other particles via the dielectric medium of empty space. The electron shell/s of every atom in Creation serves as one of a myriad number of conductive plates for the Cosmic Capacitor. So, isolation of systems of energy is an impossibility. And since the Conservation of Energy, as well as the thermodynamic laws, are all predicated upon this hypothesis of the isolation of any system of energy, the editors over at Wikipedia have admitted (on behalf of all physicists) to energy isolation being a mere hypothesis which is useful for analyzing systems of energy as a thought experiment, but does not conform to physical reality. It's also a self-contradiction to claim that an inventor's overunity device must be a perpetual motion machine which is completely cut-off from its exterior surroundings making it impossible for any source of energy to be able to run it and is thus unacceptable to the Patent Office, while thermodynamics is calculated on this same unacceptable premise of the isolation of all conventional systems of energy. Hence, it is abusive behavior to hold anyone responsible for upholding a fantasy outside of the imagination of the human mind. This is what the United States Patent Office does regarding any circuit which purports to generate more energy OUT than IN. They blanketly disregard such submissions on the grounds that they are perpetual motion machines, and thus continue to perpetuate this fraudulent misappropriation of an hypothesis to a law of physics. This is my complaint in a nutshell. Maybe the editors over at Wikipedia should be conferring with your governor friend since I'm merely citing these Wikipedia editors, along with the illogical consequences of their position?

Private Criticism #1

1. Here's a criticism, plus my response, to an email from a friend ...

My response to a friendly comment:

> "Free Energy is a myth and only fools believe they will generate Free Energy from High Voltage."
>
> When do I promote the use of HV? I don't. I promote its opposite in conjunction with an open transmission line terminated at its far end with a self-looping self-short so as to step aside and avoid the suppression of over-reactance.
>
> Or, an extremely reduced input of voltage wherein inductance plays second fiddle to capacitance, as in this example > > >
>
> Capacitance is the progenitor of Free Energy. Yet, it takes inductance to manifest it.
>
> So, even though inductance is considered to be the bane of free energy, that's 'cuz we overlook capacitance.
>
> Isolation has gone to their head. Lenz Law does not exist within a vacuum.
>
> For everything, there is an opposite.
>
> In other words, ...
> There is no free energy due to Lenz Law.
> versus...
> There *is* free energy due to Lenz Law when Lenz Law, itself, is opposed by capacitance (which is the inverse relation of Lenz Law since Lenz Law is not a primary law; it is the consequence of reactance, namely: inductive reactance).
>
> Even reactances are not laws, for they are a consequence (so, I'm told) of Maxwell's equations.
>
> Negative impedance utilizes conventional impedance as a beneficial influence instead of our continuously complaining about its existence and the inability to avoid it. Fine. No problem, man.
>
> It's ridiculous to subsist on, agree with, their pomposity of disbelief. For, they are choosing to ignore capacitive reactance when they make the Egg of Columbus claim (not to be confused with Tesla's Egg of Columbus) along with also (simultaneously) ignoring their rank isolationism is what's at the root of their disbelief. Nothing exists in isolation to anything else except as a convenient fiction. And they just happen to be the most ardent supporters of isolationism. Gee, I wonder why?
>
> All fictions require, as their predicate, a forceful entry into the innocent psyche of a child's curiosity, resulting in substituting this innocence with a new belief that: Santa Claus does not exist, nor does the Tooth Fairy. Why don't they throw out Christ while they're at it and God? Oops, sorry. They already have!
>
> Why was Christ crucified?
> 'Cuz he was a radical. He taught that God is within everyone's heart, not up in some remote sky, and that every man is his own priest. But Constantine couldn't stand that. So, he convened a council after Christ's death to eradicate the teachings of Christ by forming his own religion predicated upon a bunch of priests and make all of the priests subservient to himself and replace the teachings of Christ with Christianity. What a crock of ****.
>
> Power, reactive power, is what Free Energy is predicated upon; not energy. Energy is the congregation of day laborers who will build whatever church they are told to build by whomever tells them to build it. They are sheep (hmmm..., rhymes with: *[bleep]*).

Latest, updated, schematic drawing (https://ufile.io/il0hiucg) of a simulation (http://vinyasi.info/mhoslaw/Parametric%20Transformers/2023/Jan/Byron%20Brubaker%20Tesla%20Hairpin.zip) of Byron Brubaker's rendition of Tesla's Hairpin circuit.

But power, or I should say, the manipulation of power (the elements of reactive power) is how energy gets spent and determines its efficiency of expenditure.

Does it take a century to spend a fixed amount of energy? Or, can it be spent in a micro second? And what will be the impact of modifying its duration of expenditure?

The duration of spending a fixed quantity of energy is merely one factor of electrical reactance. And electrical reactance is merely one of two types of reactances. Magnetic reactance, which I have not studied, also exists to complicate the situation.

If it takes a minimum duration of time compression to spend a fixed quantity of energy to overcome a fixed quantity of impedance or resistance, then it's going to be a challenge to efficiently spend that fixed quantity of energy. Like trying to get past a wall of a certain height (the wall of resistance/impedance), may require a specific minimum expenditure of energy to overcome the height of that analogous wall of resistance and impedance. And if stretching time out to a greater length of duration dilutes the expenditure of energy per unit of time by too much of a factor of dilution, then there may not be enough energy per unit of time to get past resistance, and/or, impedance.

BTW, inductive impedance is the progenitor of Lenz Law (worth repeating). And, Maxwell's equations are the progenitor of the Law of Lenz.

But there are other factors of electrical reactance, not to mention magnetic reactance, which can alleviate the challenge of overcoming resistance/impedance. If I can't shorten the span of time in which a fixed quantity of energy must be spent to overcome impedance and resistance, then (at the very least) I can manipulate: capacitance, inductance, frequency of oscillations, and the phase relations between voltage and current to substitute for whatever lack of freedom I might possess (under various circumstances) of manipulating the duration of the expenditure of energy.

And by acknowledging that duration of time compression is not the only asset at my command, I acknowledge that no one is a slave to anyone else.

The reason why they push physics upon as if it were the only religion in town, is to endorse its consequential fantasy, that: Einstein's time dilation is the only game in town. Bull ****.

I met the reincarnation of Albert Einstein while I was in college three decades ago. He's a mess! He's sorry he used his fame to push for our use of nuclear power.

He spends every night until the wee hours of predawn visiting pool halls to distract himself from his private hell. He thinks he's such a fantastic pool shark as a consequence of this obsession of his to forget his past.

Well, I'm no pool shark. Yet, with a great deal of diligent focus, I managed to beat him. Me....A novice, who rarely picks up a pool stick, managed to beat a non-novice two games out of three on my first attempt. Whoops...

Ignorance is the only thing we need to fear. Everything else is a wash.

Old, Worn Out Criticisms

The following sections express an attitude which is partially true as far as *tunnel vision* can provide. They do not explain the whole story of free energy.

Old, Opening Statement

I have to concede that so-called **Free Energy** is merely the unaccountable theft of energy from outside of any system of so-called *free energy*, such as: a circuit which purports to possess an output which is greater than its input, is *actually* failing to account for the energy which it is stealing from its surroundings.

This admission leads to a conclusion that science is Godless, namely: that science refuses to admit that its precious *Conservation of Energy* is not a Law due to the non-isolation of systems of energy conceded in the Wikipedian article, entitled: *Isolated system*. Consequently, thermodynamic entropy is a lie due to the presumption which physics imposes upon our Universe that this Universe is *probably* isolated from everything else other than itself when science already concedes that *nothing has been found to be isolated* and thermodynamics *depends* upon isolation of energy systems in order to be useful for their analysis.

Just because an idea is useful doesn't make it a fact!

This makes our presumption of the *probable* isolation of our Universe a hypocritical lie.

So, :::

We're worse than Godless! We're hypocrites!

Here's another example of our Godlessness, ...

> Mathematics *officially* refuses to admit that unity (the number one) is the first and simplest prime number. In other words, mathematicians refuse to admit that the smallest number which cannot be divided by anything other than itself and the number one is also the number one! I found merely one book, forty years ago, in the mathematics section of the Powell Library at UCLA which admitted to sharing my perspective that unity is the *first* prime number.

Powell Library

> The whole problem with our definition of primality is how we word it due to worrying about whether our definition is also suitable for all higher orders of the complexity and breadth of the broad topic of number theory. These higher orders of number theory have no relevance to the simple definition of primality which should *restrict* itself to itself and extend itself no further. Because primality is a mathematical statement of a number which has no relation to any other number, ie. no equivalence to any other number, other than possessing a relationship towards itself. Hence, primality is merely a byproduct of a **self-relevant number** whose first and smallest element (of the set of self-relevant numbers) is the number 1. Divisibility, or any other *beneficial* property, is not relevant to the definition of primality. Again, **utilization is not a fact; it is merely a beneficial property** *(aka, a useful byproduct)* **of a fact**. We have lost sight of the root-causation of primality and of the root-causation of systems of energy and probably for a lot of other topics within various fields of science as well!

Thus, :::

> The divisors of the number 5 are 5 and 1 making both 5 and 1 prime by association of two equivalent types of numbers both of which belong to the same set of a pair of numbers which are the smallest divisors of the number 5. To say that the number 1 is somehow different than the number 5 is to suggest its exclusion from the divisor test for primality and further suggests that the number 5 *could be excluded as well* by some trickery of logic (aka, hypocrisy) ...

> A **prime number** (or a **prime**) is a natural number greater than 1 that is not a product of two smaller natural numbers. A natural number greater than 1 that is not prime is called a composite number. For example, 5 is prime because the only ways of writing it as a product, 1 × 5 or 5 × 1, involve 5 itself **[and the number 1]**.

> To *exclude* the number 1 from a definition of primality (in the quotation, above, from its Wikipedian article on primes) and then (in the same sentence) *include* the number 1 as participating in the *only* possible factorization of a prime number is nothing less than a self-contradiction amounting to, yet, another example of the Godlessness of science occurring in our modern world!

The smallest and simplest golden ratio is the number one found by dividing the side of an equilateral triangle by its base.

> How much more sinfully complicated need this become?

> Only when we lie do we spin complicated webs of intrigue when we should be keeping everything as simple as can be.

> Unity is also the first of an infinite series of Golden Ratios found by dividing the length of the side of the smallest odd-sided equilateral polygon, namely: a triangle, by the length of its base yielding the number one.[105]

> This refusal, on the part of mathematicians, that unity is the first and smallest prime number is like saying that the Universe does not exist as the result of the act of creation of a singular Supreme Being making mathematics another Godless branch of science.

In other words, I must do what science refuses to do. In order to accurately account for the energy of a so-called: *free energy* circuit, I must admit to our dependency upon something else which is outside of our self, namely: not equal to our self.

Hence, my overall assessment of so-called: **Free Energy** is pseudoscience since science demands hypocrisy (https://www.wordnik.com/words/hypocrisy) of itself which makes *me* non-scientific by exclusion from the act of hypocrisy!

You know what I think?

Empty space is the closest thing we'll ever get to concede as being the singular, first-cause of our Universe.

I am not finished! I am merely resting … (https://vinyasi.podbean.com/e/i-m-finished-hallelujah/)

To Allow One Standard Among Scientists and Impose Another Among the Public

::: is :::
::: beyond hypocrisy: It is deceit intended to confuse.

This deceit is executed by …
::: physicists who have postulated a mere principle of the Conservation of Energy predicated upon a fantasy of isolation to uphold a circular logic that, "there is no such thing as a free lunch," without providing any proof for this isolation. This principle of isolation is presumed to occur among all systems of energy, both large and small, for the privilege (among physicists) of analyzing their energetic pathways. It was never, or should never, have been intended to be applied *to all systems of energy all of the time as if it were a law by claiming it to be a law*, for this enslaves the mass of humanity into a status quo of ignorance and confusion over the goal of labor.

Is the purpose of labor to support our life? Or, is the purpose of life is to enjoy our labor?

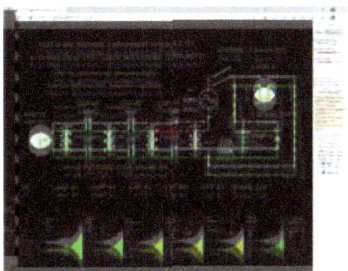

Escalating voltage differences arising from pairs of inductive and capacitive reactances in a ladder formation "… with the shunts (rungs) being inductors and the capacitors running in series (railings)."

"A circuit cannot have negative static resistance (be active) over an infinite voltage or current range, because it w[c]ould have to be able to produce [acquire] infinite power [from its environment]" over an "…unlimited range of frequencies.…"

An additional fantasy is imposed upon an unsuspecting humanity (by physicists) that a gluttony of energy **must** be force-fed *into* a circuit to ensure the circuit remains isolated from its environment, preventing any attempt the circuit may make to become self-sufficient from its environment. This self-reliance *might have occurred* had the circuit been allowed to operate under the input of a scarcity of energy. This *would have encouraged* the possibility of an over-reactance capable of moving energy *against gradients of positive impedance*. This contrary style of an unconventional movement of energy is an artifact of magnetic and electrical reactance which is given the name of: *negative impedance*.

But, :::
::: we are not allowed this optional privilege. We have created an artificial, manmade preposition of isolation among all members of society = not merely among all of our circuitry. And all of this is due to the imposition which is made upon an unsuspecting public of a principle as if it were a law when, in fact, it is merely a principle of Conservation held among cloistered physicists for the purpose of energy analysis. The public is never allowed the privilege of dreaming fantasies of their own in which energy is allowed to enter into a circuit from its environment and freely leave it under the reactive guidance of negative impedance. We're told that such a dream is in violation of the public laws of physics. Yet, no such law exists among physicists, themselves. It's merely a mental exercise among themselves while it's a public law for everyone else.

If this is not an example of deceit fostered among an elite class to maintain their status of privilege and maintain the remainder of us enslaved to confusion, then I don't know what is.

Thief or Victim of Theft?

is.gd/thieforvictim

As mortals, we are either presumed to be a thief (whether or not this is true), or else we are the victim of a presumed theft. There is no liberty for choosing neither case unless we are immortal – in which case, our immortality precludes us from having any liability imposed upon us which we do not choose to partake of.

Mortals have no choice of any kind except the choice of whether: "to do or not to do", whenever opportunity arrives and knocks on our door of conscientious attention-span. Immortals have an infinite range of choices to choose from. Predestiny is a consequence of mortality. Free-will is exclusively available to immortals. Everything else is a fiction.

As mortals, we are either sinners, or else we are the victim of someone else's sin.

It is a piece of fiction for a physicist to claim that an electrical circuit can exist in isolation and be subject to entropy ...

> An isolated system obeys the conservation law that its total energy–mass stays constant. Most often, in thermodynamics, mass and energy are treated as separately conserved.
>
> Because of the requirement of enclosure, and the near ubiquity of gravity, strictly and ideally isolated systems do not actually occur in experiments or in nature. Though very useful, they are strictly hypothetical.[106] [107] [108]
>
> Classical thermodynamics is usually presented as postulating the existence of isolated systems. It is also usually presented as the fruit of experience. Obviously, no experience has been reported of an ideally isolated system.

It is a consequential lie to claim that Current,[37] or Electric Charge,[36] can be (or, must be) Conserved. These are fictions born of the mind of man.

Perpetual motion of machines is a fiction born and sustained by the imagination of physicists. This, along with the idea of the Conservation of Energy, plus its root cause emanating from the imaginary fiction of an isolated system of energy, are all lies we tell ourselves.

> It is, however, the fruit of experience that some physical systems, including isolated ones, do seem to reach their own states of internal thermodynamic equilibrium. Classical thermodynamics postulates the existence of systems in their own states of internal thermodynamic equilibrium. This postulate is a very useful idealization. [Editor's note: In other words, "a very useful idealization" becomes a self-fulfilling proposition born of a circular argument which makes it acceptable to choose **theft over honest toil** – see, next quotation, below.]

To glamorize this particular flavor of the "fruit of experience," and to elevate this fictional ability for some isolated systems to reach their own state of equilibrium, we impose standards based on these lies in order to transform these lies into self-fulfilling errors of judgment. These standards *require* that we add the amount of energy (which will run our appliance) to its losses and supply this total assessment of energy to our appliance for it to sustain itself. This is a presumption; this is *not* a truth. The only reason it is a law is to ensure the continuation of this presumption as a self-fulfilling proposition born of a circular argument ...

> ... as Bertrand Russell observed, "The method of 'postulating' what we **want** has many advantages; they are the same as the advantages of **theft over honest toil**."[109]

In reality, circuits steal energy from every other circuit in existence or else circuits supply all of their energy requirements from sources of energy which exist outside of themselves. Either way, we get away with assuming that energy is *fictionally conserved* whether or not we *actually* account for all of it.[2]

Don't get me wrong; please don't misunderstand me. I'm not against theft, for theft is practically a *requirement* imposed upon the impoverished intended for their short-term survival in lieu of any long-range plan of extrication. What I *am against* are the lies which are used by the thief to hide his/her theft from *any* scrutiny. For *with these lies,* we become idiots.

If we lie about our sin, then we throw away our integrity. But, if we're honest about our sin, then we must have *a very good reason* for engaging in something that we know ahead of time is wrong.

Magical Me! (https://vinyasi.podbean.com/) Podcasts at Podbean

Are we the victim of theft, or are we the thief? Which are we? Because, we cannot be neither! (https://vinyasi.podbean.com/e/are-we-the-victim-of-theft-or-are-we-the-thief-which-are-we-because-we-cannot-be-neither/)

> Either we are sinners, or else we are the victims of sin. (https://vinyasi.podbean.com/e/either-we-are-sinners-or-else-we-are-the-victims-of-sin/)
>
> How many stages of transition does it take to turn a fiction into a lie? (https://vinyasi.podbean.com/e/how-many-stages-of-transition-does-it-take-to-turn-a-fiction-into-a-lie/)

Thus, I must beg to differ whenever the United States Patent Office, or any other patent office of foreign countries, claims they don't want to *look* at an inventor's submission if it purports to generate more output than its input since (they claim) that this violates the laws of physics (namely, the first and/or second law of thermodynamics plus the conservation laws of current and charge). I must protest their gross violation of logic because they are misrepresenting the inventor's device by implying that it is an isolated universe, for only an isolated universe (complete unto itself) could be an isolated system of energy, and only an isolated system of energy could possibly support the first and second laws of thermodynamics plus the conservation laws of current and charge. Anything, and I mean *anything,* inside of any universe (and, thus, comprising a mere *portion* of that universe) cannot possibly support these so-called laws of physics because this would be a violation of the definitions of these laws (predicated, such as they are, upon isolated systems of energy).

No scientist has ever managed to completely isolate an experiment. *Never!* It has never happened.

Yet, the patent office expects from the inventor what *they,* the patent office, has never accomplished, themselves!

Bogus!

The Lies We Keep Telling Ourselves Become *Truths*

It may not be possible to create an isolated system. The universe, this universe in which we reside, may not be an isolated system for the following two examples ...

Tat Wale Baba (https://search.brave.com/search?q=Tat+Wale+Baba&source=web), an Indian Yogi, was considered by his peers to be indisputably accomplished in the artistry of Vedanta in which: "I am That (Immutable, Eternal Being), thou art That, [and] all [of] this [universe] is That."[110]

While India was still under British rule (not too long ago), a British dude came to visit with Tat Wale for the purpose of becoming Tat Wale's devotee. So, to test Tat Wale's state of accomplishment, this British dude asked Tat Wale to perform a miracle to demonstrate Tat Wale's state of consciousness. So, Tat Wale put his two hands together as if he were making *namaste,* and then he drew his two hands apart. And in the space between his two hands had formed a galaxy swirling around and hanging there in the space between his two hands.

Well, ...
The British dude freaked out and ran away!

Another example, ...
When Eric Dollard (https://www.ericpdollard.com/) was a teenager, he'd meet with his friends in the garage of his parent's home for the purpose of performing their latest experiments in electricity. They had decided, on this one series of occasions, to take burnt out bulbs (which the city of San Francisco had removed from its streetlamps) and subject them to high intensity electrical oscillations while the bulb was held in the null space between the opposing poles of two coils. There, in that neutral zone, appeared within the bulb a swirling galaxy for a few seconds until the bulb couldn't take the buildup of pressure anymore and burst. And for a fraction of a second after the bulb had burst, this galaxy continued to swirl around without any boundaries of glass or whatever to hold it there before it eventually winked out.

In both of these examples, someone is pumping electromagnetic and/or electrostatic energy into each of these two galaxies to keep them alive and form them in the first place. They are *not* isolated systems. Consequently, I have to doubt that our Creator does any differently with our Universe. I seriously doubt that our universe is an isolated system of energy. So, on this

The Aura of a Yogi

The aura of a yogi is an indication of how much life force he has focused and accumulated, by way of extraction, from his environment into his body.

Upon the death of a yogi's body, this aura disperses back into the environment from which it was extracted. Hence, it could be said that a yogi has dispossessed the environment from some of its life force during that yogi's lifetime and the yogi allows the environment to repossess all of his life force upon his death.

What composes the environment of a yogi? All of the living creatures who exist within the biosphere of a yogi's environment including the Earth Spirit, Herself.

Hence, a yogi temporarily *steals* (borrows) energy, in the format of life force, from his environment until he is done using it upon his death.

And if a single, or several, devotees of this yogi are lucky enough to garner the attention of this accomplished yogi, then during that yogi's lifetime, all of these devotees will enjoy the benefit of that tremendous life force while the yogi is alive. This is called: the grace of the yogi, or in the Sanskrit it is rendered as: 'darshan'. Once that yogi passes away, all of that beneficial life force which each and every devotee of that yogi had enjoyed while their master was alive, will return to the yogi's environment and each of his devotees will suddenly become devoid of the 'grace of their master.'

If you wish, you may substitute the term of: 'parent' or 'teacher' in place of 'yogi' and replace: 'devotee' with 'child' or 'student'. They all amount to the same set of relationships.

I merely wish to point out how similar is our life by comparison to the lifespan of a 'free energy' circuit. It is not necessary to exclusively understand engineering without recourse to everyday commonplace experience if we truly understand our topic.

Tribulations

- Wikibooks: Requests for deletion – Free Energy does not Exist

    "@Leaderboard, I forgot to mention...
    Electrical engineering has always allowed for *free energy circuits* by renaming them: *unstable*. In other words, "energy IN does not equal energy OUT," defines an *unstable circuit* in which you can't predict the output based on the input, alone... *This is in addition to an allowance for a shift in time due to the frequency component of the formulae for electrical reactance supersedes Conservation of Energy.*[111] Thus, Conservation of Energy is not a law so much as it is a yardstick by which circuit topologies are measured to determine a circuit's type: If a circuit's output is unpredictable (based on its input, alone), then it is unstable since its output was not conserved within the boundaries imposed by its input. Type-casting is not disallowance; it is merely prejudice.

    "Considering how unstable circuit simulators are (due to their consistent use of matrix algebra as a shortcut for calculating a circuit's outcome), simulating an unstable circuit within the context of an unstable simulator yields "matrix is singular" error messages more often than not. Only stable circuits yield predictable outcomes. Simulators find no fault with these types of circuits.

    "Using an inherently unstable simulator to calculate a circuit's behavior is a predisposition (ie, prejudice) towards favoring stable circuits since only stable circuits will pass through this artificial, manmade act of filtration without coughing up and freezing in mid-stride. This is not due to some Law of Nature. It is due to flagrant social engineering. -- Vinyasi (discuss · contribs) 03:15, 31 October 2022 (UTC)"

- Wikibooks: Reading room, assistance – How do I improve my wikibook, or is it impossible to improve it? (https://en.wikibooks.org/wiki/Wikibooks:Reading_room/Assistance#How_do_I_improve_my_wikibook,_or_is_it_impossible_to_improve_it?)
- Wiktionary: Information desk: Two very different definitions of "perpetual motion" ... (https://en.wiktionary.org/wiki/Wiktionary:Information_desk/2022/November#Two_very_different_definitions_of_%22perpetual_motion%22_...)

Translation and Commentary of Circuit Analysis

This is an excellent opportunity for me to explain *my version* of Ohm's Law (which I introduce within the section entitled, *Block Diagram*) and how it integrates into Khan Academy's analysis of a 'live' circuit (https://www.khanacademy.org/science/electrical-engineering/ee-circuit-analysis-topic) ...

Current equals Reactive Voltage divided by Impedance ...

$$\text{Current} \equiv \left(\frac{\text{Reactive Voltage}}{\text{Impedance}} \right)$$

And ...

Watts equals the Application (the Input) of Real Voltage times its Resultant Output of Reactive Voltage divided by various Impedances (both Real and Imaginary) within a framework of time ...

$$\text{Watts/Unit of Time} \equiv \left(\frac{\text{Real Voltage Input} \times \text{Reactive Voltage Output}}{\text{Impedances}} \right) \Big/ \text{Unit of Time}$$

This is mistakenly, overly simplified into becoming *Power equals Voltage Squared Divided by Resistance*: $P \equiv \frac{V^2}{R}$ which throws away so much information that we become ignorant of what is *really* going on within our analysis of a 'live' circuit.

Unit Two, Lesson 1: Ideal Circuit Elements (https://www.khanacademy.org/science/electrical-engineering/ee-circuit-analysis-topic/circuit-elements/v/ideal-circuit-elements)

For resistors ...

$$\text{Applied Voltage} \equiv \left(\frac{\text{Reactive Voltage} \times \text{Resistance}}{\text{Impedance}} \right)$$

For capacitors ...

$$\frac{\text{Reactive Voltage}}{\text{Impedance}} = \text{Capacitance} \times \left(\frac{\text{a Change in Applied Voltage}}{\text{a Change in Time}} \right)$$

For inductors ...

$$\text{Applied Voltage} \equiv \text{Inductance} \times \left(\frac{\text{a Change in Reactive Voltage}}{\text{Impedance} \times \text{a Change in Time}} \right)$$

A few consequences of these relationships are that ...

1: Inductance *absorbs* Imaginary Reactance and converts it *into* the application of Real Voltage, while ...
2: Capacitance *generates* Imaginary Reactance by its conversion *from* the application of Real Voltage.

Alternative Explanation of *Current Reversal*

- How to Reverse Current Direction: a single page from the WikiBook, entitled: "Circuit Idea".

For Further Study

- Capacitor: Current and voltage reversal (https://en.wikipedia.org/wiki/Capacitor#Current_and_voltage_reversal)
- Reflection phase change: Electrical transmission lines (https://en.wikipedia.org/wiki/Reflection_phase_change#Electrical_transmission_lines)

broken / open line

A transmission line terminated with an open circuit is the dual case; the voltage wave is shifted by 0° and the current wave is shifted by 180°.

- Power gain
- Infinite Open-loop gain
- Inrush current & File:Inrush current.gif
- reversed time; reversed or negated damping.
- Two-port network needs a new definition for half-port networks composed of merely one terminal.
- Open-circuit voltage
- Category:Electrical parameters
- Admittance

Tesla's Magnifying Transmitter

- Audio recording (https://www.youtube.com/watch?v=AbB_l1GqEbY) of Eric Dollard reading from his essay, entitled: "Theory of Wireless Power (http://ericpdollard.com/wp-content/uploads/2018/04/theory_of_wireless_power_eric_dollard.pdf)" (1986)
- Radiant Energy is the Precursor to Free Energy (https://www.youtube.com/watch?v=jIguoTEGzyw) (a YouTube video) with parallelisms to Nikola Tesla's Wireless Transmission theory.
- Sending electricity through the Earth (https://www.youtube.com/watch?v=D3GSHRgV9SM) (a YouTube video), by Ernst Willem van den Bergh, of Wardenclyffe Research (https://www.youtube.com/user/TheMage00000) (a YouTube channel).
- Wardenclyffe (https://www.youtube.com/watch?v=bBhVDcZwAls) (YouTube video, with my comments and the OP's responses) ...

> Me ...
> *"Could the reversal of current, relative to the polarity of voltage within his Magnifying Transmitter, be a diagnostic check that his Magnifying Transmitter was succeeding at doing its job of collecting atmospheric electricity? In as much as, this reversal of current would be directing the flow of charges into his device (from the atmosphere) in contradistinction, and in counter-opposition, to conventional devices? Convention dictates that our devices must dissipate their potential to do work since they must follow the dictates of thermodynamics such that their current is in phase with their voltage with little or no separation of their phase relations (at least no separation greater than plus or minus one-quarter cycle of oscillations), and - thus- behave in an entropic manner?*
>
> *"Also to consider, is the fact that his Magnifying Transmitter was orienting its potential in a radial manner, rather than in a circulating manner, since it possessed no return path (it was a monopolar device). Thus, reversal of its current (if this had been the case) would have directed potential inwardly towards itself in the format of a flow of current directed inwardly from the surrounding environment?*
>
> *"Also, it sounds like a verification that the Ammann brothers' so-called: Atmospheric Generator may have been patterned off of Tesla's Magnifying Transmitter? Now that I've watched this video, this seems more likely than ever before since I've already considered the possibility that they were using one of his patents for their inspiration. But now, it seems very likely!*
>
> *"In further confirmation...*
> *"It was claimed, by authorities in Washington, D.C., that upon the arrival of C. Earl Ammann with his batteryless EV to demonstrate his technology for the benefit of the U.S. Patent Office, he was promptly arrested on charges of stealing energy from the grid since his demonstration in Denver, Colorado, in August of 1921 (prior to his arrival in Washington) had the distinct side-effect of putting out the power of the grid's customers in the outskirts of downtown Denver (yet, not within downtown Denver, itself). I suspect that he was messing with the phase relations of the grid across the radius of influence of his device (which he has been quoted as saying that it had a ten-mile broadcast radius). So, I'm guessing that he wasn't stealing energy from the grid so much as he was disturbing it throughout its radius of influence while at the same time supplying it with reversed polarity of the flow of energy towards the center of this range of influence at the location of his device. So, at the outskirts of this circle of influence, his device was too weak to have any influence other than that of disturbing the phase angle (or, power factor) of the grid without being strong enough to suck any current through the grid (and from the atmosphere) towards his device at this periphery*

> *of its range of influence. Thus, a more accurate assessment would be to claim that he was a domestic terrorist at the outskirts of town (if we would have created that term back then) while also being a Robin Hood of sorts within downtown Denver!"*
>
> Postscript ...
> I have found, with my six years of experience simulating overunity (over-reactive) circuits, that they will usually behave like a glutton and hog energy from a voltage source, but only if the source is provided by way of a hard electrical connection (closed switch (http://vinyasi.info/ne?startCircuit=overunity-breakthrough2.txt) in this example), or else by a "soft" connection, such as: a magnetic coupling. {Maybe this is what the Ammann brothers' device made use of? A magnetic coupling to the electric utility grid, nearby, and -thus- rightfully warrant the arrest of C. Earl Ammann as noted above?}
>
> If, on the other hand, the source is quickly disconnected, its drainage (by the overly reactive circuit) will be minimized, and this circuit topology will turn to its own reactance to make up the difference, but only if it is isolated from exterior sources, such as: the grid (which pervades the city landscape). So, one could say that sources of energy *might* get in the way of overunity circuits and, thus, block our attempt at reducing our dependency on those sources of energy (for example: the grid, batteries, solar, geothermal, etc.). Yet, sources of energy are necessary to initiate over-reactance. This is why I have learned to use precharged capacitors, of one micro volt or one milli volt mimicking environmental ambient energy at ground level, to initiate over-reactance and quickly dissipate its precharged energy into the circuit so as to avoid suppressing the evolutionary growth of over-reactance (emanating from this unique form of positive feedback).
>
> > OP (https://www.youtube.com/channel/UC9RuDKWbf05CEr6Ss7lWvUQ) ...
> > No, in my experiments I have not seen a reversal of polarity. What happens is that you receive additional current. This is also what you see in lightning. For example 20 MV is just enough to break through 20 m of air, which it does, but then another 20m step is taken, and another etc. (google "stepped leader") This builds up a charges channel and when it connects to ground it discharges violently. This final discharge contains charges collected from the atmosphere surrounding the leader (NOT - as most people assume - from the thunder cloud). As for the Ammann brothers, I have never heard of them, so I can't respond to that.
>
> Me ...
> I thought arcing spark gaps exhibit the reversal of current (relabeled: negative resistance), yet, mathematically equivalent?
>
> On a different note, and getting back to my question ...
>
> I wonder if reversal of current occurs only in the receiver coils? Not in the transmitter?
>
> > OP (https://www.youtube.com/channel/UC9RuDKWbf05CEr6Ss7lWvUQ) ...
> > Not sure if I understand what you mean. If you have a 100KHz coil resonating, the current through it reverses 200,000 times per second. In a SGTC, when the spark-gap breaks, it triggers a reversal of current in the primary, starting an oscillation.
>
> Me ...
> Then, I can assume there is more than one way to accumulate charge other than by reversal of current? And less explosive since, maybe, it is easier to ...
>
> 2:48 (https://www.youtube.com/watch?v=bBhVDcZwAls&t=168s)
> ... regulate the magnetic field drawing in charges from the atmosphere when the acceleration of the electrostatic field occurs between the cathode and the anode?
>
> > OP (https://www.youtube.com/channel/UC9RuDKWbf05CEr6Ss7lWvUQ) ...
> > Yes, I think there are more ways. In fact I am working on one that I hope will work on a much smaller scale.

Ether Theory & Gravity

- Tesla vs. Einstein: The Ether & the Birth of the New Physics (http://www.wakingtimes.com/tesla-vs-einstein-the-ether-the-birth-of-the-new-physics/) Waking Times

- Tesla's Dynamic Theory of Gravity (https://electricalscience.quora.com/Teslas-Dynamic-Theory-of-Gravity) Quora - Electrical Science

- Nikola Tesla's Dynamic Theory of Gravity (https://www.youtube.com/watch?v=sNrquwHVUPQ) (YouTube video)

- Summation of Tesla's Dynamic Theory of Gravity (https://www.netowne.com/technology/important/); An excerpt from: Occult Ether Physics (https://search.brave.com/search?q=occult+ether+physics+william+r+lyne&source=web), by William R. Lyne (https://search.brave.com/search?q=william+r+lyne&source=web).

A Few Quotations from Mahatma Gandhi

- **My life is my message.**
  - Response to a journalist's question about what his message to the world was. *Mahatma: Life of Gandhi, 1869-1948* (1968) Reel 13 (http://www.gandhiserve.org/video/mahatma/commentary13.html)
- **You assist an unjust administration most effectively by obeying its orders and decrees. An evil administration never deserves such allegiance. Allegiance to it means partaking of the evil:**
  A good person will resist an evil system with his whole soul. Disobedience of the laws of an evil state is therefore a duty.
  - *Non-Violent Resistance* - Often misquoted as "You assist an evil system most effectively by obeying its orders and decrees. An evil system never deserves such allegiance."
- All humanity is one undivided and indivisible family, and each one of us is responsible for the misdeeds of all the others. I cannot detach myself from the wickedest soul.{*citation needed* (https://en.wikipedia.org/wiki/Wikipedia:Citing_sources)}

A good person will resist an evil system with his whole soul. Disobedience of the laws of an evil state is therefore a duty.

Notes

a: ::: some of the outputs or internal states growing without bounds.

References

1. Undefined vs Indeterminate in Mathematics (https://www.cut-the-knot.org/blue/GhostCity.shtml)
2. Low Frequency Oscillations in Indian Grid (https://www.researchgate.net/publication/324978006_Low_Frequency_Oscillations_in_Indian_Grid)
3. Is it possible to obtain current indirectly from power lines? (https://skeptics.stackexchange.com/questions/3520/is-it-possible-to-obtain-current-indirectly-from-power-lines) Skeptics, StackExchange
4. What is Newton's third law? (https://www.khanacademy.org/science/physics/forces-newtons-laws/newtons-laws-of-motion/a/what-is-newtons-third-law) – Khan Academy
5. Research Gate Article (https://www.researchgate.net/publication/310445375_Why_Free_energy_is_mathematically_and_physically_possible#fullTextFileContent): Why Free energy is mathematically and physically possible, by Patrick Cornille
6. My answer to this question on Quora: *How can I make money using the science of energy?* (https://qr.ae/pridSI)
7. E_R will be a complex number (https://quantummechanics.ucsd.edu/ph130a/130_notes/node475.html#section:selfenergy), the real part of which represents an energy shift, and the imaginary part of which represents the lifetime (and energy width) of the state.
8. Nathan Stubblefield used two parallel wires, one of copper and one of iron, wrapped around the central core of his Earth Battery which he patented in 1898 (https://patentimages.storage.googleapis.com/de/5d/14/a57ffad14ccd94/US600457.pdf)
9. Nathan Stubblefield was a genius! (https://vinyasi.podbean.com/e/nathan-stubblefield-was-a-genius/)
10. Here is the original treatment of this subject. (http://vinyasi.info/patent/private/Burying%20our%20Overunity%20Circuits%20to%20Eliminate%20their%20Electrostatic%20Buildup.pdf) Its simulation files are located here (https://ufile.io/qhfx7apl) and mirrored here (http://vinyasi.info/mhoslaw/Parametric%20Transformers/2022/Dec/simplest-overunity-circuit-you-will-ever-see.zip).
11. Eli Pasternak (https://www.quora.com/profile/Eli-Pasternak) answers my question on Quora: Of what significance is the conservation of energy if its derivation from Kirchhoff's laws is assigned to a fictional concept of a node? (https://www.quora.com/Of-what-significance-is-the-conservation-of-energy-if-its-derivation-from-Kirchhoff-s-laws-is-assigned-to-a-fictional-concept-of-a-node/answer/Eli-Pasternak?ch=10&oid=402970609&share=02e1137e&srid=3zXXz&target_type=answer)
12. A question I proposed at the StackExchange physics forum as to whether I am disrupting physics by asking a question about a possible oversight (https://physics.stackexchange.com/q/741393/160958):
13. Mutual Inductance is the A Priori of Negative Resistance
14. Conservation of Energy does not Exist!
15. Paul Falstad's electronic simulator (http://falstad.com/circuit/)
16. The SPICE Page (http://bwrcs.eecs.berkeley.edu/Classes/IcBook/SPICE/)

17. Various answers to my question on Quora (https://www.quora.com/Can-a-solar-panel-tolerate-an-amperage-which-is-100x-greater-than-its-rated-voltage-So-if-its-rated-voltage-is-1V-could-it-tolerate-100A-flowing-through-it-regardless-of-the-source-of-this-amperage-And-could-it): Can a solar panel tolerate an amperage which is 100x greater than its rated voltage? So, if its rated voltage is 1V, could it tolerate 100A flowing through it regardless of the source of this amperage? And could it tolerate AC and at that amplitude?
18. Ronald Williams (https://www.quora.com/profile/Ronald-Williams-176) answers my question on Quora (https://www.quora.com/What-would-happen-if-a-reverse-matching-voltage-1-5V-of-high-current-1-5V-150A-was-sent-through-an-oscillator-fed-by-a-solar-panel-1-5V-150A/answer/Ronald-Williams-176): What would happen if a reverse matching voltage (-1.5V) of high current (-150A) was sent through an oscillator fed by a solar panel (+1.5V, -150A)?
19. Abdel Fudadin's (https://www.quora.com/profile/Abdel-Fudadin) post in Quora's Science Lounge (https://sciencelounge.quora.com/The-trick-IMHO-is-to-reduce-excessive-nuclear-power-regulation-then-get-out-of-the-way-https-www-quora-com-Wha?ch=10&oid=89367178&share=e1d988c2&srid=3zXXZ&target_type=post)
20. Free Voltage Exists! (https://vinyasi.podbean.com/e/free-voltage-exists/) = *Magical Me*, Podbean, podcast.
21. Maybe you should not believe a word I say for your own sake!? For your own protection to prevent you from harming yourself?! (https://vinyasi.podbean.com/e/maybe-you-should-not-believe-a-word-i-say-for-your-own-sake-for-your-own-protection-from-harming-yourself/) – *Magical Me*, Podbean, podcast.
22. Forgive me, for I have sinned! (https://vinyasi.podbean.com/e/forgive-me-for-i-have-sinned/) – *Magical Me*, Podbean, podcast.
23. Impedance is a Source of Energy Reactive Voltage (https://electricalscience.quora.com/Impedance-is-a-Source-of-Energy-as-its-Conclusion-Update-After-we-ve-managed-to-generate-voltag) – Quora
24. Voltage Performs Work (https://www.facebook.com/photo.php?fbid=5818999488122857&set=b.5818999488122857&type=3) = This is what Stanley Meyer was all about.
25. What does "negative impedance" mean in electricity and electronics? Has the capacitor negative impedance? How do we create a "negative" capacitor? (https://www.researchgate.net/post/What_does_negative_impedance_mean_in_electricity_and_electronics_Has_the_capacitor_negative_impedance_How_do_we_create_a_negative_capacitor) by Cyril Mechkev (https://www.researchgate.net/profile/Cyril-Mechkev)
26. A Free-Energy slideshow (https://www.youtube.com/playlist?list=PLwteYqUouDFCg3vTIsor_HXxHSBJbFBQj) set to a backdrop of popular tunes
27. A YouTube mix of Alexandre Desplat's music beginning with: *The Heroic Weather-Conditions of the Universe, Parts 1-7* (https://www.youtube.com/watch?v=ycNCAsIpZF4&list=RDycNCAsIpZF4&start_radio=1&rv=ycNCAsIpZF4&t=0) = right-click to open in a new tab
28. Chopin Nocturnes (https://www.youtube.com/watch?v=-gDinVAmtA0) = Wonders Of Classical Music (https://www.youtube.com/channel/UC6MVyFIWtb3NJR4LUvttbMQ)
29. Ravi Shankar = Classical Sitarist (https://www.youtube.com/watch?v=Ik60OonblOk&list=PLwteYqUouDFDiChq7w8Q_GstH1zUnqskg)
30. Indian Sitar, Instrumental Music, 10 Hours (https://www.youtube.com/watch?v=D6B4xo6zYdk)
31. Christmas Songs (https://www.youtube.com/playlist?list=PLwteYqUouDFBL3ZkB7uiVsnTxlw5uQMDj)
32. People of the Condor and Eagle (https://www.ioes.ucla.edu/news/23832/), by Dangung Dennis (https://www.ioes.ucla.edu/person/dangung-dennis/), 2017 Pritzker Emerging Environmental Genius Award finalist.
33. An Ancient Legend Meets Modern Times: The Eagle and the Condor Prophecy (https://threadsofperu.com/blogs/blog/an-ancient-legend-meets-modern-times-the-eagle-and-the-condor) – Threads of Peru Blog
34. How can one show that imaginary numbers really do exist? (https://www.math.toronto.edu/mathnet/answers/imaexist.html) This is an argument that imaginary numbers exist authored by the Mathematics Network of the University of Toronto.

Editor's note: Their only success is that they prove the concept of imaginary numbers is a valid fantasy whose validity lies solely in its self-consistency. They also do not prove any analog in the real world which mirrors this imagination existing within the minds of mathematicians. In fact, they admit that there need not be any analog in the physical world in order to have validity in the world of imaginary mathematics. Thus, nothing relevant to a *physical proof* of imaginary numbers has been offered.

This is important, because it is upon this frail basis that the United States Patent Office refuses to peruse any application for patent which purports to export more energy than it imports.

How can the Patent Office have any authority if it utilizes faulty logic?
Answer...
Obviously, a preference is being exercised which favors convention (and, possibly, vested interests) over reason!

35. Chapter 11 – Parametric Amplifiers and Oscillator (https://www.nii.ac.jp/qis/first-quantum/e/forStudents/lecture/pdf/noise/chapter11.pdf) from First Quantum Information Lecture Series (https://www.nii.ac.jp/qis/first-quantum/e/forStudents/lecture/)
36. Kirchhoff's Voltage Law (http://www.electronics-tutorials.ws/dccircuits/kirchhoffs-voltage-law.html) = Electronics Tutorials, DC Circuits.
37. Kirchhoff's Current Law (http://www.electronics-tutorials.ws/dccircuits/kirchhoffs-current-law.html) = Electronics Tutorials, DC Circuits.
38. US patent office reveals number of secret patents (https://web.archive.org/web/20090831012621/https://www.newscientist.com/blogs/shortsharpscience/2008/10/us-patent-office-reveals-numbe.html)
39. "Physics on the Fringe (https://think.kera.org/2011/12/07/alternative-theories-of-everything/): Smoke Rings, Circlons, and Alternative Theories of Everything," by Margaret Wertheim (Walker & Company, 2011):

40. Free energy, imaginary numbers and electrical reactance all exist in time apart from space. (https://vinyasi.podbean.com/e/free-energy-imaginary-numbers-and-electrical-reactance-all-exist-in-time-apart-from-space/)
41. Taking the square root of a negative number on faith is the predicate of electrical engineering! (https://forum.allaboutcircuits.com/threads/can-someone-explain-the-behavior-of-this-transformer.190225/#post-1776905)
42. Is anyone able to explain Eric Dollard's concepts of space and counter-space? (https://www.quora.com/Is-anyone-able-to-explain-Eric-Dollards-concepts-of-space-and-counter-space/answer/George-Mardari) Quora
43. Jeffrey Denenberg's (https://www.quora.com/profile/Jeffrey-Denenberg) answer to: When an open transmission line is terminated by a shorted transmission line, do they produce a purely imaginary impedance at their input? Can this reactance grow at exponential rates if input is kept extremely small and restricted to a single moment? (https://www.quora.com/When-an-open-transmission-line-is-terminated-by-a-shorted-transmission-line-do-they-produce-a-purely-imaginary-impedance-at-their-input-Can-this-reactance-grow-at-exponential-rates-if-input-is-kept-extremely-small/answer/Jeffrey-Denenberg?__filter__=all&__nsrc__=notif_page&__sncid__=33092740614&__snid3__=44479886759) on Quora
44. Reflections cause several undesirable effects, including modifying frequency responses, causing overload power in transmitters and overvoltages on power lines. *However, the reflection phenomenon can also be made use of in such devices as stubs and impedance transformers.* The special cases of open circuit and short circuit lines are of particular relevance to stubs.
45. Zip compressed file (https://ufile.io/3pbinpmr) of three Micro-Cap simulated variations of this phenomenon. All of them produce similar results despite the use of three galvanic-style batteries in one version versus not in the other two versions. Peruse this directory (http://vinyasi.info/mhoslaw/Parametric%20Transformers/2022/Nov/?C=M;O=D) on my website (as an alternative to downloading this ZIP compressed file) and hunt for any filename which begins with *simplest-overunity-circuit-you-will-ever-see__*.
46. Is it Possible to Generate Current without Voltage? (http://vinyasi.info/patent/pri-vate/Burying%20our%20Overunity%20Circuits%20to%20Eliminate%20their%20Electrostatic%20Buildup.pdf) – (posted to my website)
47. The Oversight of the Ammann Brothers' (https://fuel-efficient-vehicles.org/energy-news/?page_id=971) Fuel Efficient Vehicle

> "While Earl was demonstrating his invention all over the streets of Denver, the power had been cut off in the foothills. In spite of this, when he went to Washington DC shortly afterward to try to obtain a patent on his Cosmo Electric Generator, he found that charges had been filed against him claiming he had a device to steal power from the power lines."
>
> K. H. Isselstein,
>
> Spokane, WA

48. Pentagon Aliens, Chapter VIII: A Taste of Other Energy Secrets (https://www.bibliotecapleyades.net/ciencia/pentagonaliens/pentagonaliens08.htm) – Quote: *"... for every 200 pounds of iron connected to the device, a full horsepower was added to it."*
49. An example of negative resistance (http://vinyasi.info/ne?startCircuit=negresist.txt) simulated in Paul Falstad's circuit simulator (mirrored copy).
50. Jim Phipps answer (on Quora) to: With closed magnetic coupling between primary and secondary, what will improve in a transformer? (https://www.quora.com/With-closed-magnetic-coupling-between-primary-and-secondary-what-will-improve-in-a-transformer/answer/Jim-Phipps-1)
51. Antenna Ground Plane: (https://www.electronics-notes.com/articles/antennas-propagation/grounding-earthing/antenna-ground-plane-theory-design.php) Theory & Design » Electronics Notes (electronics-notes.com)
52. The ground plane is not merely a zero reference. (https://vinyasi.podbean.com/e/the-ground-plane-is-not-merely-a-zero-reference/)
53. My discussion of Byron's Hairpin Circuit (https://electricalscience.quora.com/I-took-Byron-Brubakers-Hairpin-circuit-https-www-schematics-com-project-10-power-factor-hairpin-vic-pump-tesla-br) on Quora
54. Micro-Cap 12 electronic simulator (http://www.spectrum-soft.com/index.shtm)
55. LTspice – Fast • Free • Unlimited (https://www.analog.com/en/design-center/design-tools-and-calculators/ltspice-simulator.html)
56. glossary definition of *relative permittivity* (https://glossary.ametsoc.org/wiki/Relative_permittivity)
57. Department of Physics, Aerosol Refractive Index Archive, category: rocks and conglomerates, substance: granite. (http://eodg.atm.ox.ac.uk/ARIA/data?Rocks_and_Conglomerates/Granite_(Toon_et_al._1977)/granite_Toon_1977.ri)
58. For comparison with other dielectric substances, see: Georgia State University's *Hyperphysics* webpage on the "Dielectric Constants at 20°C." (http://hyperphysics.phy-astr.gsu.edu/hbase/Tables/diel.html)
59. Toppr question (https://www.toppr.com/ask/en-us/question/the-capacitance-of-two-concentric-spherical-shells-of-radii-r1/)
60. Utility Companies Defy Speed of Light by placing Capacitors on Transmission Lines Every 100 Miles (https://www.youtube.com/watch?v=h05TfwBfuhw&feature=youtu.be)
61. What is the typical voltage for power lines? (https://www.hunker.com/13418990/what-is-the-typical-voltage-for-power-lines)
62. The Types of Transmission Lines Based on Voltage (https://resources.system-analysis.cadence.com/blog/msa2021-the-types-of-transmission-lines-based-on-voltage)
63. mirrored copy of: Cloudbusting UFOs @ Vimeo (https://vimeo.com/vinyasi/cloudbustingufos)
64. mirrored copy of: How to Crash a UFO @ Vimeo (https://vimeo.com/vinyasi/crashaufo)
65. United States Patent No. US 10,144,532 B2; Dec. 4, 2018 for neutralizing inertial mass (https://patentimages.storage.googleapis.com/de/4c/43/62c585ccc936cc/US10144532.pdf)

66. UFOlogy: A Major Breakthrough in the Scientific Understanding of Unidentified Flying Objects, (https://a.co/d/8qNIfAq) Paperback – September 1, 1976, by James M. McCampbell.
67. Pages: 77 and 78 of chapter 13, entitled: *1973 – A Key Sighting*, in a book by David Alzofon, entitled: *Gravity Control with Present Technology.* (https://a.co/d/eEpZDzc)
68. Anti-Gravity with Present Technology: Implementation and Theoretical Foundation (http://www.tuks.nl/pdf/Reference_Material/Reactionless%20Drive/Alzofon%20-%20Anti-Gravity%20wth%20Present%20Technology%20-%20Implementation%20and%20Theoretical%20Foundation%20-%201981.pdf)
69. Dynamic nuclear orientation: C. D. Jeffries, (Wiley, London, 1963. viii-177, p. 45 s). (https://doi.org/10.1016/0029-5582(64)90380-3)
70. Did Nikola Tesla invent a car that ran without fuel? (https://www.quora.com/Did-Nikola-Tesla-invent-a-car-that-ran-without-fuel/answer/Vin-Yasi)
71. Everything You Need to Know About the Battery in Your Car or Truck (https://www.batteriesplus.com/blog/power/car-battery-care): What Happens When Your Battery's Charge Gets Too Low?, by Bryan Veldboom @Batteries Plus
72. This is the first stage of my simulated development (http://vinyasi.info/mhoslaw/Parametric%20Transformers/2022/Sept/ammann%20with%20solar%20capacitance,%20v2c3c,%20schematic.png) of the notion that maybe the inspiration for the Ammann brothers' Cosmic Atmospheric Generator came from the spark-transmitter of Heinrich Hertz? It is located within this directory (http://vinyasi.info/mhoslaw/Parametric%20Transformers/2022/Sept/) on my website. (http://vinyasi.info/)
73. Homemade diodes (https://overunity.com/612/homemade-diodes/) plus their tutorial (https://learn.sparkfun.com/tutorials/diodes). — *"Darn! Which end is the cathode?"*
74. Borax or Baking Soda Rectifier and the glow. (http://www.sparkbangbuzz.com/els/borax-el.htm)
75. What Is ESR and Why Does It Matter? Part 1 (https://www.skeletontech.com/skeleton-blog/what-is-esr)
76. Through power factor correction, using a capacitor in parallel with an inductive load, we can reuse 99% of our electricity in this example (http://vinyasi.info/ne?startCircuit=powerfactor2.txt). This spawns *the appearance of* a 100 to 1 gain of output relative to input. Yet, this *appearance* is a mirage since no law of physics has been violated.
77. Search terms: simulation round off error (https://search.brave.com/search?q=simulation+round+off+error&source=web)
78. I have discovered that round-off error is a byproduct of our choice of computer which we use for hosting our simulation software. Is it a 64-bit computer? Then, we're in luck! Is it anything less than this? Oops! BTW... Round-off error is *not* determined by the simulator, per se. I wrote about this, a little bit with illustrated examples, in my self-published book, entitled: *Oops! How I Goofed Simulating Overunity Circuits on a 32-bit Computer...,* available at Amazon (https://a.co/d/2LQxdXr), Payhip (https://payhip.com/b/0zqG4) and for free download from my website (http://vinyasi.info/mhoslaw/Oops.pdf).
79. Erik Anson answers a question on Quora: Is inertia also a force, like gravity, but the opposite? (https://www.quora.com/Is-inertia-also-a-force-like-gravity-but-the-opposite/answer/Erik-Anson)
80. My answer (on Quora) to the question: Has anyone tried to recreate Joseph Newman's perpetual motion machine? (https://www.quora.com/Has-anyone-tried-to-recreate-Joseph-Newmans-perpetual-motion-machine/answer/Vin-Yasi)
81. Eric Dollard's Analog Computer as a Power Amplifier (https://electricalscience.quora.com/Eric-Dollard-s-Analog-Computer-as-a-Power-Amplifier)
82. This text: *The Moon's Rotation* (http://teslacollection.com/tesla_articles/1919/electrical_experimenter/nikola_tesla/the_moon_s_rotation) is read aloud by a narrator (on YouTube), entitled: *The Moon's Rotation* (https://www.youtube.com/watch?v=ipZEhIpjbG8) ♦ By Nikola Tesla ♦ Physics & Mechanics ♦ Audiobook.
83. Mechanical–electrical analogies: Classes of analogy
84. A few answers (on Quora) to the question of: What would happen if an induced current did not oppose the change that caused it, as in Lenz's law? (https://www.quora.com/What-would-happen-if-an-induced-current-did-not-oppose-the-change-that-caused-it-as-in-Lenzs-law)
85. Please see: Tesla's invention of the Vacuum (tube) Capacitor. The shortcut URL for this Wikipedia article, is: https :// is.gd / teslacap
86. Newton's First Law of: *Inertia.*
87. The alternative to remanence (preserving magnetism) is capacitance (to retard electrostatic potential) (https://www.youtube.com/watch?t=3h45m15s&v=cCJcU7INwnU&feature=youtu.be). The shortcut URL for this video excerpt, is: https :// is.gd / spacetimeconjunction

88: Translation from the Russian into English of a Youtube comment #1 (https://translate.google.com/?sl=auto&tl=en&text=...&op=translate) on MrPreva's video (https://www.youtube.com/watch?v=XlnN3jk1Hy0).

89. Translation from the Russian into English of another Youtube comment #2 (https://translate.google.com/?sl=auto&tl=en&text=%D0%9B%D1%83%D1%87%D1%88%D0%B5%20%D1%80%D0%B0%D0%B1%D0%BE%D1%82%D0%B0%D1%82%D1%8C%20%D0%B2%20%D1%81%D0%B2%D0%BE%D0%B5%D0%BC%20%D0%BA%D0%B0%D0%BA%20%D0%B0%D0%BA%D0%BA%D1%83%D0%BC%D1%83%D0%BB%D1%8F%D1%82%D0%BE%D1%80%20%D1%81%D0%BE%20%D1%81%D0%B2%D0%BE%D0%B5%D0%B9%20%D0%BF%D0%BE%D0%BB%D0%BD%D0%BE%D0%B9%20%D0%BE%D0%B1%D1%80%D0%B0%D1%82%D0%BD%D0%BE%D0%B9%20%D1%81%D0%B2%D1%8F%D0%B7%D1%8C%D1%8E%2C%20%D0%B2%D0%BA%D0%BB%D1%8E%D1%87%D0%B0%D1%8F%20%D0%B2%D0%BA%D0%BB%D1%8E%D1%87%D0%B5%D0%BD%D0%B8%D0%B5%20%D1%81%D0%B2%D0%BE%D0%B5%D0%B3%D0%BE%20%D1%81%D0%BE%D0%B1%D1%81%D1%82%D0%B2%D0%B5%D0%BD%D0%BD%D0%BE%D0%B3%D0%BE%20%D0%B2%D1%8B%D1%85%D0%BE%D0%B4%D0%B0%20%D0%B2%D0%BE%20%D0%B2%D1%85%D0%BE%D0%B4%2C%20%D0%B0%D0%BD%D0%B0%D0%BB%D0%BE%D0%B3%D0%B8%D1%87%D0%BD%D0%BE%20%D1%81%D1%85%D0%B5%D0%BC%D0%B5%20%D0%9D%D1%8C%D1%8E%D0%BC%D0%B0%D0%BD%D0%B0.&op=translate) on MrPreva's video (https://www.youtube.com/watch?v=XInN3jk1Hy0).

90. Can someone explain the behavior of this transformer? (https://forum.allaboutcircuits.com/threads/can-someone-explain-the-behavior-of-this-transformer.190225/) – All About Circuits Forum

91. Schematic Slideshow (https://josephnewman.info/schematics) of LTSPICE screenshots simulating Newman's device

92. A shorted transformer could yield over-unity. But is it power? Or is it just free-spinning? (https://vinyasi.podbean.com/e/shorted-transformer-yielding-over-unity-but-is-it-power-or-is-it-just-free-spinning/)

93. Jim Murray's Transforming Generator (http://vinyasi.info/energy/shorted-transforming-generator.mp3) – This is an audio excerpt from his video presentation (https://search.brave.com/search?q=Jim+Murray+Transforming+Generator&source=web)

94. Search Bing for: "PMH perpetual motion holder edward leedskalnin (https://www.bing.com/search?q=PMH%20perpetual%20motion%20holder%20edward%20leedskalnin&qs=n&form=QBRE&=%25eManage%20Your%20Search%20History=%25E&sp=-1&pq=pmh%20perpetual%20motion%20holder%20edward%20pe%20leedskalnin&sc=10-46&sk=&cvid=28BA0BC94E54E4DA88E1176BEC5BF90&ghsh=0&ghacc=0&ghpl=)"

95. 2002 Toyota Rav4 EV (http://evnut.com/rav.htm)

96. A idea of the First Principle (https://www.quora.com/A-idea-of-the-First-Principle-a-priori-first-cause-postulate-axiom-primitive-notion-etc-is-founded-on-the-proposition-that-there-are-self-evident-propositions-But-what-is-the-first-principle-upon-which-this-very?ch=10&oid=1020736636&share=c3203b28&srid=3zXXZ&target_type=question) (a priori, first cause, postulate, axiom, primitive notion, etc.) is founded on the proposition that there are "self-evident propositions." But what is the "first principle" upon which this very "proposition" is founded?

97. Some Recent Developments of Regenerative Circuits (https://play.google.com/books/reader?id=bNI1AQAAMAAJ&pg=GBS.PA344&hl=en) by Edwin H. Armstrong; Proceedings of The Institute of Radio Engineers, vol: 10; 1922

98. Physics paper (http://www.physicsessays.org/browse-journal-2/product/677-17-frederick-e-alzofon-the-unity-of-nature-and-the-search-for-a-unified-field-theory.html)

99. Anti-Gravity with Present Technology: Implementation and Theoretical Foundation, F.E. Alzofon, 1981 (http://www.tuks.nl/pdf/Reference_Material/Reactionless%20Drive/Alzofon%20-%20Anti-Gravity%20with%20Present%20Technology%20-%20Implementation%20and%20Theoretical%20Foundation%20-%201981.pdf) and a backup copy (http://vinyasi.info/circuits1/texts/Frederick%20Alzofon/Alzofon%20-%20Anti-Gravity%20with%20Present%20Technology%20-%20Implementation%20and%20Theoretical%20Foundation%20-%201981.pdf)

100. Gravity Control With Present Technology (https://emediapress.com/shop/gravity-control-present-technology/#:~:text=In%201960%2C%20and%20an%20Air%20Force%20survey%20of%20gravitation,derived%20from%20the%20model%20of%20the%20gravitational%20force.) = audio/video presentation at Emedia Press

101. My answer to: Now that a majority understand green energy can't meet demand and is the main cause of inflation driven by ideology not science, will we see more nuclear power? (https://www.quora.com/Now-that-a-majority-understand-green-energy-can-t-meet-demand-and-is-the-main-cause-of-inflation-driven-by-ideology-not-science-will-we-see-more-nuclear-power/answer/Vin-Yasi?ch=10&oid=401301739&share=8c7141a78&srid=3zXXZ&target_type=answer) on Quora

102. My answer to: What would happen if a reverse matching voltage (-1.5V) of high current (-150A) was sent through an oscillator fed by a solar panel (±1.5V, -150A)? (https://www.quora.com/What-would-happen-if-a-reverse-matching-voltage-1-5V-of-high-current-150A-was-sent-through-an-oscillator-fed-by-a-solar-panel-1-5V-150A/answer/Ronald-Williams-176?ch=10&oid=403424417&share=401b6cd1&srid=3zXXZ&target_type=answer) on Quora

103. A Facebook commentary by Rachel Kocsis (https://www.facebook.com/vinyasi/posts/pfbid02fp17ubyjc8MpDwaPQ2cEdNVqdLBTrChAqeZYNPdD9JwByfseWLZaRHPjf8M3FWSDI?comment_id=828944958400792)
104. The Eternal Duties of a Human Beings (https://bhagavad-gita.org/Gita/verse-03-24.html)
105. The Infinite Range of Golden Ratios (http://vinyasi.info/Infinite%20Range%20of%20Golden%20Ratios/) – archived (https://web.archive.org/web/20221103021206/http://vinyasi.info/Infinite%20Range%20of%20Golden%20Ratios/)
106. Thermodynamics of Spontaneous and Non-Spontaneous Processes; (https://books.google.com/books?id=2RzE2pCfijYC&pg=PA3) I. M. Kolesnikov et al, pg 136.
107. A System and Its Surroundings; (http://chemwiki.ucdavis.edu/Physical_Chemistry/Thermodynamics/A_System_And_Its_Surroundings#Isolated_System) UC Davis ChemWiki, by University of California – Davis.
108. Hyperphysics, (http://hyperphysics.phy-astr.gsu.edu/hbase/conser.html#isosys) by the Department of Physics and Astronomy of Georgia State University.
109. Feinberg, Joel; Shafer-Landau, Russ (2008). *Reason and responsibility: readings in some basic problems of philosophy*. Cengage Learning. pp. 257–58. ISBN 9780495094920.
110. All This Is That, The Beach Boys (https://genius.com/The-beach-boys-all-this-is-that-lyrics) – lyrics.
111. Each cycle of oscillation is keeping time for an electrical reactance to continue to occur. If this frequency should change of its own accord, then time has shifted within the domain of that reactance and Conservation is disqualified (under Noether's Theorem).

Translations

- Español (https://en-wikiversity-org.translate.goog/wiki/Draft:Free_Energy_does_not_Exist?_x_tr_sl=en&_x_tr_tl=es&_x_tr_hl=en&_x_tr_pto=wapp)
- Deutsch (https://en-wikiversity-org.translate.goog/wiki/Draft:Free_Energy_does_not_Exist?_x_tr_sl=en&_x_tr_tl=de&_x_tr_hl=en&_x_tr_pto=wapp)
- Francés (https://en-wikiversity-org.translate.goog/wiki/Draft:Free_Energy_does_not_Exist?_x_tr_sl=en&_x_tr_tl=fr&_x_tr_hl=en&_x_tr_pto=wapp)
- Italiano (https://en-wikiversity-org.translate.goog/wiki/Draft:Free_Energy_does_not_Exist?_x_tr_sl=en&_x_tr_tl=it&_x_tr_hl=en&_x_tr_pto=wapp)
- Português (https://en-wikiversity-org.translate.goog/wiki/Draft:Free_Energy_does_not_Exist?_x_tr_sl=en&_x_tr_tl=pt&_x_tr_hl=en&_x_tr_pto=wapp)
- Polski (https://en-wikiversity-org.translate.goog/wiki/Draft:Free_Energy_does_not_Exist?_x_tr_sl=en&_x_tr_tl=pl&_x_tr_hl=en&_x_tr_pto=wapp)
- Русский (https://en-wikiversity-org.translate.goog/wiki/Draft:Free_Energy_does_not_Exist?_x_tr_sl=en&_x_tr_tl=ru&_x_tr_hl=en&_x_tr_pto=wapp)
- 中國, 台灣 (https://en-wikiversity-org.translate.goog/wiki/Draft:Free_Energy_does_not_Exist?_x_tr_sl=en&_x_tr_tl=zh-TW&_x_tr_hl=en&_x_tr_pto=wapp)
- 中国人, 中国大陆 (https://en-wikiversity-org.translate.goog/wiki/Draft:Free_Energy_does_not_Exist?_x_tr_sl=en&_x_tr_tl=zh-CN&_x_tr_hl=en&_x_tr_pto=wapp)
- 日本 (https://en-wikiversity-org.translate.goog/wiki/Draft:Free_Energy_does_not_Exist?_x_tr_sl=en&_x_tr_tl=ja&_x_tr_hl=en&_x_tr_pto=wapp)
- हिन्दी (https://en-wikiversity-org.translate.goog/wiki/Draft:Free_Energy_does_not_Exist?_x_tr_sl=en&_x_tr_tl=hi&_x_tr_hl=en&_x_tr_pto=wapp)
- 한국인 (https://en-wikiversity-org.translate.goog/wiki/Draft:Free_Energy_does_not_Exist?_x_tr_sl=en&_x_tr_tl=ko&_x_tr_hl=en&_x_tr_pto=wapp)
- عربي (https://en-wikiversity-org.translate.goog/wiki/Draft:Free_Energy_does_not_Exist?_x_tr_sl=en&_x_tr_tl=ar&_x_tr_hl=en&_x_tr_pto=wapp)
- فارسی (https://en-wikiversity-org.translate.goog/wiki/Draft:Free_Energy_does_not_Exist?_x_tr_sl=en&_x_tr_tl=fa&_x_tr_hl=en&_x_tr_pto=wapp)

Shortcuts to this Page

Desktop
https :// is . gd / draftfree
https :// is . gd / freedraft

Mobile
https :// is . gd / draftfreem
https :// is . gd / freedraftm

Retrieved from "https://en.wikiversity.org/w/index.php?title=Draft:Free_Energy_does_not_Exist&oldid=2475863"

This page was last edited on 22 February 2023, at 18:36.

Text is available under the Creative Commons Attribution-ShareAlike License; additional terms may apply. By using this site, you agree to the Terms of Use and Privacy Policy.

www.ingramcontent.com/pod-product-compliance
Lightning Source LLC
Chambersburg PA
CBHW051918210526
45473CB00006B/2055